Translational Research in Genetics and Genomics

Translational Research in Genetics and Genomics

Moyra Smith, MD, PhD, MFA

2008

OXFORD
UNIVERSITY PRESS

Oxford University Press, Inc., publishes works that further
Oxford University's objective of excellence
in research, scholarship, and education.

Oxford   New York
Auckland   Cape Town   Dar es Salaam   Hong Kong   Karachi
Kuala Lumpur   Madrid   Melbourne   Mexico City   Nairobi
New Delhi   Shanghai   Taipei   Toronto

With offices in
Argentina   Austria   Brazil   Chile   Czech Republic   France   Greece
Guatemala   Hungary   Italy   Japan   Poland   Portugal   Singapore
South Korea   Switzerland   Thailand   Turkey   Ukraine   Vietnam

Published by Oxford University Press, Inc.
198 Madison Avenue, New York, New York 10016

www.oup.com

Oxford is a registered trademark of Oxford University Press

Library of Congress Cataloging-in-Publication Data

Smith, Moyra.
Translational research in genetics and genomics / Moyra Smith.
    p. ; cm.
Includes bibliographical references and index.
ISBN: 978-0-19-531376-5
1. Medical genetics.   2. Genomics,
[DNLM: 1. Genetics, Medical.   2. Genetic Diseases,
Inborn.   3. Genomics. QZ 50 S647 2008]   I. Title.
RB155.S55   2008
616'.042—dc22   2007037252

9 8 7 6 5 4 3 2 1

Printed in the United States of America
on acid-free paper

To wrest from nature the secrets which have perplexed philosophers in all ages, to track to their sources the causes of disease, to correlate the vast stores of knowledge, that they may be quickly available for the prevention and cure of disease—these are our ambitions. To carefully observe the phenomena of life in all its phases normal and perverted, to make perfect the most difficult of all the arts, the art of observation, to call to aid the science of experimentation, to cultivate the reasoning faculty so as to be able to know the true from the false—these are our methods. To prevent disease, to relieve suffering and to heal the sick—this is our work.

—William Osler, 1906

PREFACE

The underlying premise of this book is that discovery of the fundamental mechanisms of a specific disease leads to improved diagnostic methods and treatment, and in some cases points the way to disease prevention. The goal of this book is to review progress in the translation of research in molecular genetics to improved diagnosis and therapy of single-gene disorders and of disorders where genetic factors play an important role.

The study of human genomics, the architecture, composition, and function of chromosomes and their segregation and transmission from one generation to the next, has steadily progressed, particularly over the last 40 years, when techniques evolved to map human genes and genetic markers. Mapping studies and positional cloning enabled identification and characterization of disease genes. Availability of DNA sequence information, through efforts in the Human Genome Project, has greatly expedited this process. More recently, sequence information, characterization of genetic variation and map locations of many thousands of genetic markers, and high-throughput analysis technologies have provided resources to approach the problems of complex common diseases where genetic risk contributes to the etiopathogenesis. There are a number of different hypotheses regarding the genetic factors involved in these diseases. I explore several of these and present results of published studies in support of each hypothesis.

Research on the human genome, facilitated by the availability of rapid sequencing and microarray technologies, has produced a number of surprising results. One example is the revelation of the high degree of structural variation in the genome that is present in healthy individuals. However, we have also come to learn that specific structural variants predispose to genome rearrangements and to dosage imbalances that may lead directly to pathology or predispose to pathology, in some cases through modification of gene–environment interactions. Furthermore, information gained from analysis of DNA sequences and RNA transcripts requires that we now redefine our long-standing definition of a gene.

Dissections of the molecular basis of single-gene disorders have led to the development of new therapies that specifically target the gene defects or the downstream effects of the gene mutation. One striking example is the finding of the high frequency of splicing defects in single-gene disorders and the identification of therapeutic agents that suppress these. Another important development is the finding that specific chemical agents may suppress the effects of nonsense mutations that generate stop codons. In a number of disorders, molecular chaperones can mitigate the downstream effects of mutations that lead to aberrant protein folding and altered protein stability. Knowledge of the molecular defects in cancer is actively translated to develop specific therapies that target malignant cells and defects in specific signaling and metabolic pathways.

Availability of DNA sequence information and high-throughput technologies have greatly impacted pharmacogenomics and drug design. Along with this progress, pharmacogenetic studies continue to reveal that individual variation in drug response needs to be taken into account. Fortunately, technical advances facilitate analysis of the molecular basis of adverse drug responses in many cases.

As late-onset neurodegenerative disease and cancer contribute more frequently to overall disease burden, it becomes even more important to consider environmental factors that impact the genome structure and function. The roles of reactive oxygen species in generating DNA damage, somatic mutation, altered protein structure, and mitochondria are important considerations. In addition, the effects of synthetic toxic substances on the genome require vigorous investigation.

The final chapter of this book is devoted to the phenome, the analysis of phenotypic complexity, and clinical services to patients. In considering who should provide genetic services, it is clear that the team should be multidisciplinary and that genetics will not remain the prerogative of regional or academic genetic centers. All branches of medicine will need to use genetic knowledge in their practice.

To some, the progress of translation of research in genetics and genomics to patient care seems remarkable; to others it seems slow and inadequate. As Lewis Thomas wrote in 1974 in *The Lives of a Cell*, "There is no question about our good intentions in this matter: we all hanker, collectively, to become applied scientists as soon as we can, overnight if possible. It takes some doing, however. Everyone forgets how long and hard the work must be before the really important applications become applicable."

ACKNOWLEDGMENTS

I thank William Lamsback and Nancy Wolitzer, editors at Oxford University Press for their guidance with this book. I wish to express my gratitude to Sheryl Rowe and Jean Blackburn for their valuable assistance in the copyediting and final stages of preparation for publication.

I have benefited greatly from the resources available through the University of California Library system, and I am grateful for these.

I wish to acknowledge the encouragement of my colleagues in the Department of Pediatrics at the University of California, Irvine.

CONTENTS

Translational Research in Genetics and Genomics

1

ISOLATION OF DISEASE GENES
THROUGH GENOMICS

Connecting phenotype with genotype is the fundamental aim of genetics.

—D. Botstein and N. Risch, 2003

Advances in chromosome mapping and DNA sequencing have greatly increased our ability to locate and characterize genes responsible for specific diseases. Physical maps of genes on chromosomes provide information on the position of a gene or marker on a chromosome relative to landmarks on that chromosome (e.g., p or q arm, chromosome band). Since completion of the human genome projects, the physical map location of a gene may also be based on defining the position of a gene or marker relative to the nucleotide sequence of the chromosome to which it maps.

Linkage maps are derived through analysis of the cosegregation of a series of polymorphic markers during meiosis and the transmission of a series of polymorphic genetic markers from one generation to the next.

The possibility of finding a genetic marker that is closely linked to a disease gene is greater if there is a high density of polymorphic markers on each chromosome, and if the markers are highly informative, that is, having many different alleles. When highly informative markers are analyzed in a population with random mating, there is a high chance that each parent will be heterozygous for each allele. Methods to identify different alleles at a specific locus based on analysis of DNA sequence have yielded markers that are more readily analyzed and that are more informative. Various types of sequence variation constitute the basis for allele identification. DNA sequence variation may lead to differences in sensitivity to DNA cleavage by a particular restriction endonuclease. Through sequencing it may be possible to identify specific nucleotide changes that give rise to a polymorphism. Knowledge of the specific nucleotide changes that give rise to a polymorphism may be utilized in the design of oligonucleotide probes to be used in microarrays. Intensity of hybridization of a labeled sample DNA fragment to the microarray will depend on the degree of match to the microarray segment. Polymorphic markers include variation in the number of nucleotide repeats in a specific DNA segment or single nucleotide differences.

In the case of rare recessive phenotypes that occur in population isolates or in inbred populations, homozygosity mapping is very useful. This form of mapping is based on the fact that affected members of a family with a specific genetic disease also inherit a specific gene region. Alleles of polymorphic markers that map in this region are identical in the affected family members.

In rare cases, mapping of a disease gene locus may be facilitated by the presence of chromosome abnormalities, including deletions, duplications, and translocations. Identification of the putative map position of a disease-determining gene on the basis of linkage analysis or chromosome breakpoint analysis represents the first step in positional cloning and isolation of the disease gene. Key elements in disease gene isolation are accurate assessment of the phenotype and accurate mapping. Phenotypic heterogeneity, where mutations in different genes give rise to the same phenotype, complicates the process.

Following mapping, database resources may be used to determine which genes map in that region and to consider the biological and physiological relevance of these genes to the disease process. The latter constitute candidate genes for the disease. In many cases, however, particularly prior to comprehensive genome sequencing, the disease-causing gene was unknown prior to its positional cloning and the underlying biological mechanism of the disease was unknown. Following fine mapping of the locus, isolation of that gene is possible.

TRANSLATIONAL ASPECTS OF GENE MAPPING AND ISOLATION

Characterization of the product of the disease gene leads to understanding of the molecular basis of diseases and elucidation of pathogenesis. Through analysis of gene product function and the role of that product in biochemical and physiological pathways, and through identification of other gene products that participate in those same pathways, it may be possible to identify genes that play a role in the etiology of similar disease phenotypes.

Identification of the disease-determining gene generates material that facilitates diagnosis. Identification of the gene product and elucidation of the pathogenic pathway may facilitate development of specific therapies.

Linkage studies have proven less powerful in the isolation of genes involved in common complex diseases. Linkage disequilibrium studies are more likely to be useful in identification of genes for these disorders (Botstein and Risch 2003). Before discussing linkage disequilibrium mapping and association studies, it will be useful to consider theories regarding the origin of common genetically complex, multifactorial diseases.

COMPLEX MULTIFACTORIAL DISEASES

Theories Regarding Origin

One theory centers on the concept that DNA changes that predispose to complex multifactorial diseases are evolutionarily old, that they therefore exist at high frequency in the populations, and that a complex disease phenotype results from interaction between genetic risk factors and specific environmental factors (Risch and Merikangas 1996). It is this theory that has fueled the multibillion-dollar search for genetic variation (polymorphisms, often single nucleotide polymorphisms, SNPs) associated with or in linkage disequilibrium with common diseases.

Ropers (2007) presented evidence that the complexity of multifactorial diseases is most likely due to genetic heterogeneity. Furthermore, he emphasized that new mutations, including those that lead to de novo copy number variation, likely contribute to the incidence of common complex diseases. Ropers emphasized further that complex disorders are not necessarily multifactorial. In some cases, one or a few genetic changes may lead to the disorder. These disorders are therefore complex in the sense that mutations in any number of different genes may lead to the disease phenotype.

Linkage Disequilibrium and Association Studies in Mapping of Complex Diseases

Linkage disequilibrium mapping is based on searches for association between disease and a specific allele at a marker locus. Furthermore, it is based on the assumption that in a particular family, affected members will inherit the same allele at disease-associated marker loci.

DNA sequence-based polymorphisms

Among the goals of projects designed to develop a detailed catalog of DNA sequence-based polymorphisms in the human genome was to have available a platform to analyze the association and cosegregation of specific DNA variants with specific common diseases. The guiding principle was that through definition and mapping of disease-associated markers, it would be possible to identify new disease pathways. This in turn would lead to improved risk assessment and open the way for development of more effective disease-specific therapies.

Association studies are currently based largely on the analysis of SNPs. In many studies, a genome-wide analysis is carried out, and this includes analysis of SNPs in coding and in noncoding regions of the genome. One of the key steps in further analysis of association is to determine the functional significance of the nucleotide variation in a SNP. Botstein and Risch (2003) proposed that the analysis of SNPs in genes is most important in association assessment. They proposed that following initial identification of an associated SNP in a gene, resequencing be carried out to identify additional and possible less polymorphic SNPs in the coding region. Furthermore, they proposed that if coding region SNPs are analyzed, it is possible to include information on the significance of nucleotide changes that have been developed over the past several decades through the analysis of mutant proteins in diseases due to single gene defects.

Development of a genetic map of approximately 8000 highly informative microsatellite markers facilitated disease gene mapping within regions of 1 to 2 million base pairs (Weissenbach 1998). Development of the map of SNPs facilitated gene mapping by linkage studies since hundreds of thousands of these markers can be analyzed using microarrays containing bound oligonucleotide probes in a cost-effective manner (Gunderson et al. 2005).

Linkage disequilibrium and HapMap

Since polymorphisms in neighboring segments of DNA often travel together, the question arose to what extent it is possible to predict one polymorphism on the basis of the presence of another. Research carried out as part of the HapMap project led to generation of information on the association of specific DNA polymorphisms in different populations. This project also sought to determine the minimum number of SNPs that would be representative of each specific haplotype. These were designated as Tag SNPs.

Haplotypes

A haplotype is a series of linked genetic markers that usually segregates as a unit during meiosis. Haplotype blocks occur as a result of blocks of linkage disequilibrium.

If a specific haplotype (i.e., a series of polymorphisms) is associated with a specific gene variant, that gives rise to a disease, the presence of a specific haplotype in an at-risk individual may serve as an indicator of the presence of the disease-predisposing mutation before the symptoms and signs of the disease are evident.

Certain mutations may be difficult to detect with certainty, particularly in samples provided for prenatal diagnosis. Knowledge of the haplotype on which a disease mutation occurs may be helpful if the haplotype can be analyzed more readily than the disease mutation.

Linkage disequilibrium

Polymorphisms in adjacent segments of DNA often travel together. Reich et al. (2001) defined linkage disequilibrium as "the correlations among neighboring alleles that reflect haplotypes descended from a single ancestral chromosome" (p. 199).

In their analysis of a U.S. population, Reich et al. (2001) reported that linkage disequilibrium typically extends for a distance of 60 kb from a common allele. They noted that in an African population (Nigerian), linkage disequilibrium extends over much shorter distances.

In constructing linkage disequilibrium maps, Reich et al. (2001) focused on core SNPs within genes and the SNPs chosen because they demonstrated a high frequency of variation. They resequenced DNA segments of about 2 kb that were located at distance of 0.5, 10, 20, 40, 80, and 160 kb from a core SNP. Their data indicated that linkage disequilibrium extends further than had been estimated prior to their study. They noted that long-range linkage disequilibrium is best explained by a founder effect or by a bottleneck in population growth. Shorter regions of linkage disequilibrium, an average of 5 kb, occurred in African populations, indicating the more ancient origin of these populations. Other populations likely arose from migrations of subsets of individuals from Africa.

Reich et al. (2001) drew attention to the implications of their study for disease gene mapping. The large blocks of linkage disequilibrium in Europeans would facilitate initial mapping of a disease gene–associated SNP. The drawback of the large blocks of linkage disequilibrium is that follow-up fine mapping would be more difficult.

Tag SNPs

Since there are strong correlations between particular variants within regions of linkage disequilibrium, and within a specific haplotype, it is possible to use one SNP (a Tag SNP) to predict the presence of other SNPs. It is therefore possible to reduce the number of SNPs that need to be analyzed in the study of genetic variants associated with specific diseases (Goldstein and Cavalleri 2005). Estimates are that through use of Tag SNPs the number of genotypes required for mapping of disease associations will be reduced by one-tenth. The HapMap project sought to discover association between specific SNP variants within haplotypes using DNA samples derived from four different populations.

In a number of studies, genome-wide SNP analysis is carried out using SNPs located in genes and in intergenic regions. In other studies, analysis is restricted to SNPs located within genes, or SNPs in regions that show a high degree of conservation in different species.

Follow-up studies indicate that Tag SNPs selected in the HapMap data transfer well to different populations. However, in African populations supplemental SNPs are useful in association studies.

HapMap: Blocklike structure of genome

Based on genotyping of more than 1 million SNPs in 269 samples from four different populations, the International HapMap Consortium (2005) confirmed a blocklike structure in the human genome. Blocks of low haplotype diversity exist; in these blocks the alleles in different SNPs are highly correlated, indicating that recombination and mutation are infrequent. Recombination hotspots occur, and these give rise to high levels of linkage disequilibrium. Linkage disequilibrium is low near telomeres and is elevated near centromeres since recombination is suppressed in the vicinity of the centromere.

Heteromorphology in the genome, due to deletions or duplications of small and large regions, turns out to be much greater than previously expected (Rocha et al. 2006). These investigators estimated that each individual harbors hundreds of copy number variations.

Based on analysis of transcript levels, SNPs, and copy number variations (CNVs) in individuals in the HapMap project, Stranger et al. (2007) determined that SNPs captured 83.6% and CNVs captured 17.7% of the variation in gene expression. They noted that both forms of variation need to be taken into account in analyses of the genome for disease association.

TRANSLATIONAL ASPECTS OF SNP AND HAPLOTYPE MAPS: COMMON DISEASES WITH GENETIC RISK

One motivation for development of the HapMap and extensive resources on SNPs was to make available reagents to search for variants associated with common diseases and to thereby identify disease pathways. One benefit of these discoveries will be improved diagnostics. Another potential benefit will be improved pharmacotherapy and especially more effective therapy for different subgroups of patients (O'Rahilly and Wareham 2006). Important advances in the identification of common disease-associated variants have been made in inflammatory bowel disease, age-related macular degeneration, and identification of a common variant.

Progress has been made in identifying genes important in diabetes mellitus type 1 and type 2 and in defining genes important in multiple sclerosis and childhood asthma.

Type 1 Diabetes Mellitus

The occurrence of diabetes is influenced by environmental factors and genetic factors. The risk in first-degree relatives of a patient with type 1 diabetes is 15 times greater than the population risk. Evidence that specific genes play a role derives from genetic linkage studies, association studies, and parent-to-child transmission disequilibrium tests. A number of different studies have shown that the major histocompatibility locus *MHCII* that encodes the leukocyte antigens HLADQ and HLADR is a major genetic determinant of type 1 diabetes (Cucca et al. 2001). Linkage studies and association studies have revealed that the insulin encoding locus (*INS*) on chromosome 11p15 plays an important role (Bennett et al. 1995). Another contributory locus to the etiology of this disorder is the *PTPN22* locus (protein tyrosine phosphatase) on chromosome 1p13 (Smyth et al. 2004).

Hakonarson et al. (2007) carried out a two-stage analysis of approximately 530 SNP markers in 563 pediatric patients with diabetes and 1146 controls. These investigators subsequently carried out an independent study of transmission of alleles from parents to children. Both studies confirmed prior associations of type 1 diabetes with *MHCII*, *INS*, and *PTPN22* loci. In addition, they identified a significant association of the disorder with a variant on chromosome 16p13 in the *KIAA0350* locus. This locus encodes an as yet uncharacterized gene. Hakonarson et al. noted that the structure of the protein predicted from the *KIAA0350* sequence indicates that it likely encodes a sugar-binding type C lectin.

Disease Gene Identified Following Genome-Wide Association Studies: Type 2 Diabetes Mellitus

A number of advances in the identification of genes that play roles in common diseases have emerged through studies in the Icelandic population. Studies of microsatellite polymorphic markers in patients with type 2 diabetes mellitus (T2DM) in the Icelandic population by Grant et al. (2006) revealed a striking association with a specific marker *DG10S478*. Six alleles occur in this marker; alleles designated 0, 8, and 12 account for 98% of chromosomes. Allele 0 has a protective association with T2DM. Grant et al. carried out genotyping of SNPs in the region of *DG10S478*. Their studies revealed a near perfect correlation between allele G in SNP *rs 12255372* and the 0 allele of *DG10S478*. Other alleles of *DG10S478* correlated with the T allele of SNP *12255372*. Further analysis proved that it was possible to collapse all the nonzero (non-0) alleles of *DG10S478* into a composite haplotype, designated X. The frequency of X in controls is 27.6 and in T2DM patients it is 36.4. They reported that heterozygous carriers of the risk alleles have a 1.45 relative risk (RR) for T2DM while homozygotes have a 2.41 RR.

Grant et al. (2006) noted that *DG10S478* occurs in a haplotype block on chromosome 10q25.2 that includes part of introns 3, exon 4, and part of introns 4 of the gene Transcription factor 7-like 2 (*TCF7L2*).

There is now evidence from different population groups, European, Asian, and African, that *TCF7L2* plays a role in T2DM. Sladek et al. (2007) carried out a two-stage genome-wide association study to identify risk loci for T2DM. The first stage involved studies on 1363 cases and controls. Important factors included in the selection of cases were age of onset of diabetes prior to 45 years and family history of one first-degree affected relative. Obese subjects were excluded; obesity was defined on the basis of a body mass index. Controls were selected on the basis of a normal fasting blood glucose level. Two array platforms were used for analysis. These included HapMap Tag SNPs (317,503 SNPs) and the Illumina Infinium Bead arrays; these include 109,365 SNPs in a gene-centered design. Together in these analyses 392,935 unique loci were examined.

The first stage of the analysis yielded 59 significant SNPs. Eight of these were in the *TCF7L2* locus that encodes a transcription factor.

In the second stage of the analysis, a different sample was analyzed, including 2617 T2DM and 2894 controls. In this stage, analysis of the 59 SNPs that showed significant association in Stage 1 were analyzed. Results revealed significant associations for eight SNPs that represented five loci. The most highly significant association was demonstrated for an SNP within *SLC30A8*, a gene that encodes a zinc transporter that expresses in the secretory vesicles of beta cells in the pancreas and plays a role in insulin biosynthesis. Other significant associations were found in a linkage block that included the insulin degrading enzyme, the homeodomain protein HHEX that plays a role in pancreatic and hepatic development, and the *KIF11* gene that encodes kinesin interacting factor 11 (KIF11).

Significant association between type 2 diabetes and the transcription factor *TCF7L2* was previously described by Damcott et al. (2006) in the Amish population. Saxena et al. (2006) reported that SNPs in *TCF7L2* are associated T2DM and with insulin response to glucose in nondiabetic individuals.

Common disease, common variant

Sladek et al. (2007) noted that for three of the four associated loci, the major allele constituted the risk allele. They considered this finding to be in support of the hypothesis that ancestral alleles were adapted to the environment of ancient populations. Today, in a different environment, these alleles increase disease risk. Their analysis did not reveal epistasis (interaction) between the loci to produce the effect. Taken together, these loci account for 70% of the population-attributable risk for T2DM in the population studied.

Age-Related Macular Degeneration

Macular degeneration (MD) is a disease that impacts the central retina. Initially it leads to distortion of fine detail perception and later central vision is completely lost. MD may occur in younger individuals as a manifestation of a monogenic disease, for example, in Stargardt vitelliform macular dystrophy and Sorsby fundus dystrophy. Genes for these diseases have been cloned. Rattner and Nathans (2006) reviewed the pathophysiology and genetics of macular dystrophy. They noted that age-related macular dystrophy (AMD) is the leading cause of blindness in adults in industrialized societies. In populations of European origin over 50 years of age, the incidence of AMD is

1.5%. In individuals older than 75 years, the incidence is 10%. The strongest risk factors for AMD are family history, age, and smoking. Smoking may increase risks through oxidative damage or vascular damage. Antioxidants such as beta-carotenes, vitamin C, and vitamin E reduce this damage.

Early indications of AMD are abnormalities of the retinal pigment epithelium, including areas of hypopigmentation and hyperpigmentation. In addition, lesions referred to as *drusen* appear. These are poorly demarcated deposits of lipid and protein. Drusen components include lipids, oxidized lipids, and proteins including ubiquitin, crystallins, immunoglobulins, and complement. In late AMD there are local regions of retinal pigment epithelium (RPE) loss and new growth of blood vessels into the retina. These vessels may rupture, leading to hemorrhages.

Rattner and Nathans (2006) reviewed animal models of AMD. Studies in mouse models revealed the importance of oxidative damage and hyperlipidemia in the pathogenesis of AMD. Mice that are deficient in superoxide dismutase ($Sod1^{-/-}$) start to accumulate drusen at 6 months of age. Exposure to white light, equivalent to sunlight, increases drusen formation.

Genes involved in age related macular degeneration: Complement pathway

The first genetic linkage study that provided evidence for mapping of AMD on chromosome 1q25-q31 was carried out in a three-generation family where this disorder segregated as an autosomal dominant trait. When age-dependent penetrance was considered, a lod score of 3.20 was obtained between AMD and microsatellite polymorphic markers (Klein et al. 1998).

Haines et al. (2005) undertook analysis of SNPs located throughout 24 Mb segments of DNA in the linked region. Studies were carried out in two different data sets. The first included samples from 182 families; the second included 495 AMD cases and 185 controls. These investigators identified significant associations between AMD and two SNPs, rs2019724 and rs6428379, that are located approximately 263 kb apart. They then identified a five-SNP haplotype, GAGGT, that occurred in 46% of cases and 33% of controls. In the family-based study, the same haplotype was transmitted to affected individuals more frequently than to unaffected individuals (significance $p = 0.0003$). They noted that the complement factor H gene and five other genes map within the region of the haplotype. Complement factor H was a candidate gene of special interest in view of its role in immune response and because inflammatory changes are prominent in AMD.

Haines et al. (2005) then sequenced the coding region of complement factor H in 24 cases and controls homozygous for the GAGGT haplotype.

STEPS IN THE DISCOVERY OF A GENE THAT PLAYS AN IMPORTANT
ROLE IN ONE FORM OF AMD.

1. GENOME-WIDE LINKAGE STUDIES PROVIDE EVIDENCE FOR OLIGOGENIC
PATTERNS OF INHERITANCE FOR AMD.

TWO SEPARATE STUDIES (WEEKS ET AL. 2004; IYENGAR ET AL. 2004)
PROVIDE EVIDENCE FOR AN AMD-DETERMINING GENE IN THE 1Q32 REGION.

2. ANALYSES OF A HAPLOTYPE SET OF 5 SNP MARKERS THAT SPAN 261 KB IN
CHROMOSOME 1Q32 (HAINES ET AL. 2005) YIELD THE FOLLOWING RESULT:

SNP ID	ALLELE	FREQUENCY OF THIS HAPLOTYPE	
		CASES	CONTROLS
rs 1831281	G		
rs 3753395	A		
rs 1853883	G	46%	33%
rs 10494745	G		
rs 6428279	T		

3. ONE GENE WITHIN THIS 261 KB REGION ENCODES COMPLEMENT FACTOR
H, IS A CANDIDATE GENE. INTRAGENIC SNP VARIANTS WERE
SOUGHT (HAINES ET AL. 2005).

LOCATION	SNP ID	AMINO ACID CHANGE	FREQUENCY IN	
			AMD	CONTROLS
EXON 9	rs 1061170	Y402H	94%	46%
			45/48	22/48

Figure 1–1.   Genome-wide linkage studies, haplotype analyses, and association studies established that complement factor H plays an important role in one form of age-related macular degeneration (AMD).

They identified one sequence variant that was more common in cases than in controls, the Y402 variant (genotype *T1227C*). This variant has functional consequences: the substitution leads to increased affinity of CFH for the complement protein C3B (Figure 1–1).

A number of studies have replicated the association of complement factor H and AMD. Maller et al. (2006) carried out additional screening of AMD patients using a denser SNP coverage of the gene encoding this factor. In their study, the strongest association was between AMD and an intronic SNP rs1410996. This SNP is in strong linkage disequilibrium with the Y420 SNP.

They determined that individuals homozygous for the high-risk haplotype had a 15 times greater risk of developing AMD than individuals in the general population.

It is of interest to note that variants in loci that encode other components of the complement pathway are also associated with AMD. Gold et al. (2006) screened loci that encode the BF and C2 regulatory proteins in the complement pathway in 900 individuals with AMD and in 400 controls. The complement factor B (BF) and complement C2 genes are located in the major histocompatibility complex III region. They identified a statistically significant common risk haplotype and two protective haplotypes. Gold et al. (2006) noted that their study provides additional evidence for the role of the alternative complement pathway in the pathobiology of AMD.

A number of different studies have led to identification of an AMD risk allele on chromosome 10 in an anonymous coding DNA segment LOC387715. The allele variant changes the coding sequence at position 69, A69S (Maller et al. 2006; Schaumberg et al. 2007). The function of the LOC397715 protein is not known.

Based on analysis of all three loci (CFH, LOC387715, and C2-CFB), and including evidence for an additive effect of variants at the 3 loci, Maller et al. (2006) determined that 10% of the population have a 40-fold increased risk and 1% have a 250-fold increased risk of AMD. The latter group has the high-risk allele at each of the three loci. They calculated that the variants at the three loci explain approximately half of the excess risk of AMD to siblings. In their study, there was no evidence of association of risk alleles at one of the loci with a specific type of AMD.

Maller et al. (2006) noted that the findings in AMD are consistent with the concepts of quantitative genetics published by R. A. Fisher in 1918. He postulated that genotypes at a small number of loci might combine resulting in a continuous risk gradient in the population.

Schaumberg et al. (2007) determined that the percentage of cases of AMD attributable to the complement factor H variant and the LOC387715 variant was 58% to 68%. They noted further that in subjects homozygous for both risk alleles, the AMD risk increased 50-fold.

Translational aspects of genetic studies in age-related macular degeneration: Treatment

Given the importance of complement regulatory proteins in the genetic predisposition to AMD, Rattner and Nathans (2006) reviewed possibilities for therapy aimed at decreasing complement activation. They postulated that the complex homeostatic network that is disrupted in AMD may be amenable to pharmacological manipulation. They noted that complement activation is mediated by proteolysis and protease inhibitors and suggested that it may be possible to discover high-affinity ligands that inhibit the relevant proteases. They further noted that activation of the immune system and complement pathways may be important in susceptible individuals. There is evidence that antibody titers to *Chlamydia pneumoniae* are higher in affected individuals (Guymer and Robman 2007). Another treatment strategy suggested by Rattner and Nathans is the use of compounds to decrease the production of transretinal, produced by the photoreceptor outer segment. They emphasize that in the development of therapy programs it will be important to stratify participants on the basis of genetic analysis.

RNA inhibition (RNAi) therapy has been applied in macular degeneration to reduce the growth of blood vessels into the eye (Dykxhoorn and Lieberman 2006).

Genes Involved in Inflammatory Bowel Disease, Crohn's Disease, and Ulcerative Colitis

Gaya et al. (2006) reviewed progress in characterization of genes that play a role in inflammatory bowel disease (IBD). The main forms of IBD are Crohn's disease and ulcerative colitis. They noted that twin studies and family studies provided evidence that inherited factors play a role in the pathogenesis of Crohn's disease and ulcerative colitis. Studies in Europe revealed that in monozygotic twins the concordance rate for these diseases is 36%, while in dizygotic twins it is 4%. The incidence of Crohn's disease is higher in family members of patients than in the general population, and the risk is greater in siblings than in other family members. The family prevalence is higher in early-onset disease than in disease of later onset. There are ethnic differences in the prevalence of Crohn's disease. The literature review carried out by Gaya et al. (2006) revealed that nine different chromosomes carried genes that play a role in IBD, based on replicated studies The locus on chromosome 16q, initially described by Hugot et al. (1996), is associated primarily with Crohn's disease. A chromosome 12 locus (IBD2) is associated primarily with ulcerative colitis.

Hugot et al. (1996) recruited families with multiple affected members with Crohn's disease. They analyzed microsatellite polymorphic markers and carried out sib-pair linkage analysis. Two independent sample panels were assessed. Their study provided evidence for linkage of Crohn's disease to chromosome 16. Subsequently, Hugot et al. (2003) carried out additional fine mapping of the chromosome 16 locus and linkage disequilibrium analysis of SNPs in a series of coding and noncoding DNA segments. They determined the SNPs in the *NOD2* gene played a role in Crohn's disease.

Crohn's disease: Gene isolation

Hampe et al. (2001) selected *NOD2* as a candidate gene in the Crohn's disease–linked region on chromosome 16 based on its function; the *NOD2* gene product *NFkappa B* activation plays a role in response to bacterial lipopolysaccharides. The *NOD2* gene subsequently became known as *CARD15*.

Crohn's disease susceptibility is associated with two common missense mutations and with a frameshift mutation in *CARD15*. These allelic variants play a less significant role in susceptibility to Crohn's disease in the Northern European population than in other European populations. In Ashkenazi Jewish populations, *CARD15* plays an important role; however, mutations other than those first identified by Hugot et al. (2001) and Hampe et al. (2001) are involved in Crohn's disease susceptibility. *CARD15* mutations are associated primarily with ileal disease and with early-onset disease.

Analysis of the molecular structure revealed that other proteins share the three domains of CARD15 protein. The leucine-rich domain is most commonly mutated in patients with Crohn's disease.

CARD15 likely acts as an intracellular sensor for specific bacterial components (Gaya et al. 2006). It is expressed in monocytes and in intestinal epithelial cells. Highest expression occurs in the Paneth cells. These cells also express defensins. Wehkamp and Harder (2004) demonstrated that *CARD15* mutations lead to decreased defensin production. Probiotics that increase defensin production have proven useful in the treatment of Crohn's disease. Probiotics are defined as "live organisms that when ingested in adequate amounts exert a health benefit" (Quigley and Flourie 2007). Gaya et al. (2006) noted that there is evidence that *CARD15* plays a role in the killing of pathogens in epithelial cells.

Based on their studies on knockout mice, Watanabe et al. (2005) reported that *CARD15* mutations lead to enhanced production of proinflammatory cytokines. They noted specifically that *CARD15* regulates *NFkappa B* activation in response to activation of the Toll receptor system including TLR2.

Importance of immune response in etiology of Crohn's disease

Gaya et al. (2006) concluded that the *CARD15* discovery has highlighted the importance of the role of the immune response to specific components in the etiology of Crohn's disease. Other *CARD* genes that play a role in IBD include the *CARD4* gene.

Toll receptors in inflammatory bowel disease

There is increasing evidence for involvement of the Toll receptors in IBD. Toll receptors are expressed on macrophages, where they facilitate phagocytosis of microorganisms. The Toll 4 receptor encoded by TRL4 is also expressed on epithelial cells, and specific polymorphisms are associated with IBD (Gaya et al. 2006).

The TRL5 receptor specifically interacts with the pathogen-associated molecule flagellin. A TLR5 polymorphism that acts as a dominant negative mutation and reduces expression of the TLR5 protein protects against Crohn's disease. This observation has prompted the suggestion that Crohn's disease may be amenable to treatment by substances that block TLR5 (Netea et al. 2005). It is of interest to note that CARD15 and TLR5 signaling are linked. Begue et al. (2006) reported that the interaction of bacterial flagellin with TLR5 regulates the production of CARD15.

There is increasing evidence that IBD is caused by an abnormal response to normal gut flora (Marx 2007). Of key importance in IBD and other auto-immune diseases is a newly discovered T lymphocyte, TH17. The activity of this T cell is influenced by a specific cytokine, interleukin 23 (IL23). Mannon et al. (2004) reported that an antibody developed against IL12 is effective in patients with IBD. Of particular interest is the fact that this antibody is also reactive against IL23, since IL23 and IL12 share a common subunit.

Interleukin 23 receptor in ileal Crohn's disease

In 2006, Duerr et al. published results of a genome-wide study to search for SNPs associated with Crohn's disease. SNPs within the IL23 receptor (IL23R) showed the strongest association. In the associated polymorphism, glutamine was substituted for arginine. The glutamine allele at position 381 in IL23R is much less common in patients than the arginine allele. The glutamine allele is protective against Crohn's disease.

Cardon (2006) emphasized that success stories in association studies in complex disease may be in part due to the way in which the phenotype is defined. Further successful studies are often carried out in stages. Duerr et al. (2006) focused on patients with ileal Crohn's disease in their first genome-wide screen. They then followed up with a screen of 401 patients and controls. Following this, they carried out family-based association studies. These studies revealed significant distortion of allele transmission.

COMPLEX DISEASES: THE CASE FOR MULTIPLE
RARE ALLELES AND INTERACTING GENES

Ropers (2007) noted that in the majority of association studies in complex diseases, reproducible associations were found in cases where very high density SNP analysis was carried out on very large cohort sizes. Nevertheless, the factors identified made a small contribution to genetic risk.

Ropers reviewed evidence that single gene defects and structural genomic changes, including copy number changes and translocations at a specific chromosome location, have played a key role in identifying important loci in the origin of complex common diseases, including mental retardation, late-onset neurodegenerative diseases such as Parkinson disease and Alzheimer disease, and psychiatric disorders such as schizophrenia.

Complex Common Diseases: Interacting Genes in a Pathway

There is growing evidence that specific common diseases may be due to mutations in genes that interact in a common pathway. In schizophrenia, the *DISC1* gene (disrupted in schizophrenia) was identified as a locus disrupted as the result of a translation involving chromosome 1 in a patient with schizophrenia. There is now evidence of the importance of *DISC1* in this disorder from a number of different studies. Genome-wide scan of Finnish patients with this disorder identified *DISC1* as a candidate gene (Hennah et al. 2007). Subsequent studies led to identification of a specific risk allelic haplotype spanning exons 1 and 2 in *DISC1*. To identify additional genes important in schizophrenia, Hennah et al. stratified the data. They analyzed individuals with the *DISC* risk allele separately from those without this allele. In the cohort that was positive for the *DISC* risk allele, they identified additional linked loci. One of these yielding a significant lod score of 3.17 was located on chromosome 16p13 and contained the *NDE1* locus (homolog of Nude mitotic phosphoprotein). This locus encodes a protein that binds to the *DISC1* gene product. Hennah and coworkers then analyzed seven SNPs in *NDE1* and identified a specific haplotype that was present in female patients with schizophrenia in their sample. These findings indicate that two interacting proteins, DISC1 and NDE1, work together to cause the disorder, at least in a subset of patients. It is interesting to note that these two proteins form a complex that functions in neuronal migration. The *RLN* gene product reelin regulates this complex upstream. *RLN* is a schizophrenia candidate gene locus.

Evidence for Extreme Heterogeneity Within Candidate Genes for Complex Diseases

Studies on an allergic condition characterized by icthyosis, eczema, and asthma revealed strong association between this condition and two null alleles in the filaggrin gene (Sandilands et al. 2007). These two null alleles were relatively frequent in the population studies. Subsequent sequencing of the gene in cases and controls led to identification of 15 rare alleles that play a role in the etiology of the disease. Sandilands et al. reported that these alleles were nonsense or frameshift mutations. They noted that SNP analyses and transmission studies would likely have failed to detect these alleles.

Rare Alleles in Three Different Genes That Play a Role in Atherosclerosis

Cohen et al. (2004) sequenced three candidate genes that contribute to quantitative variation in levels of plasma high-density lipoprotein cholesterol HDL-C. They determined that nonsynonymous sequence variants were significantly more common in individuals with low HDL-C than in those with high HDL-C. Biochemical studies revealed that the nonsynonymous variants were functionally significant. These investigators concluded that rare alleles with major phenotypic effects contribute significantly to low HDL-C levels. Studies of both men and women worldwide have demonstrated that the risk for atherosclerotic disease is inversely related to blood levels of HDL-C; that is, the higher the HDL-C, the lower the risk.

Identification of Oligogenic Loci in Complex Diseases

There is evidence that in some cases relatively few genes, each with a significant effect, contribute to the origin of complex common diseases.

Genome-wide association studies in multiple sclerosis

Hafler et al. (2007) reported results of a case control study that involved analyses in 12,360 individuals and genotyping of 334,023 SNPs. Their results revealed that multiple SNPs in the *HLADRA* locus were strongly associated with multiple sclerosis, $p = 8.94 \times 10^{-81}$. There was also evidence, though much less significant, of association of multiple sclerosis with SNP in the interleukin 2 receptor, alpha IL2RA 2.96×10^{-8} and with the interleukin 7 receptor 2.94×10^{-7}.

In reviewing these results, Peltonen (2007) drew attention to the fact that previous small-scale studies have demonstrated the importance of the *HLADR* locus, and that previous linkage studies revealed linkage of multiple sclerosis to the *IL2RA* locus. Peltonen noted that these findings make the case for full allelic examination of candidate genes identified in genome-wide linkage studies.

Genes important in the etiology of childhood asthma

Moffatt et al. (2007) reported results of studies in 994 patients with childhood asthma and 1243 controls, using arrays with 317,000 SNP markers. They demonstrated that multiple polymorphic markers on chromosome 17q21 were associated with the disorder. In a subsequent replication study on a different population, the same chromosome region was found to be associated with the diagnosis of childhood asthma. No other chromosome region showed significance at the 1% cutoff in the false discovery rate method of determination of statistical significance.

In lymphoblastoid cell lines from patients, the expression analysis of 14 genes located within was then carried out. Moffatt et al. determined that expression of a specific gene correlated with the SNP allele. This gene, *ORMDL3*, encodes a transmembrane protein that is anchored in the endoplasmic reticulum.

2

ADVANCES IN GENOME ANALYSIS: DISCOVERY OF STRUCTURAL VARIATION AND GENOMIC DOSAGE DIFFERENCES IN HEALTH AND DISEASE

Resources to search for genomic variation have rapidly expanded with availability of the genomic sequence, development of fluorescence-based genomic hybridization in analysis of chromosomes, and particularly through availability of microarray technologies. Comparative genomic hybridization using arrayed Bac clones with large inserts (approximately 170 kb) of human genomic DNA has been particularly informative. Structural genomic information is also becoming more intensely studied through use of oligonucleotide arrays and single nucleotide polymorphism (SNP) arrays (described below).

One of the most surprising discoveries in human genome research between 2005 and 2007 relates to the extent of copy number variation (CNV) of large genomic segments. In establishing the role of CNVs in genetic disease, it is particularly important to distinguish between normal variations and disease-causing variations. The application of extended SNP data and of microarray data to analysis of human genetic disease requires that background information be gathered to determine the extent of normal variation in the genome in that population.

Thus identification of CNV in a subject with a particular disease requires analysis in the unaffected parents and siblings, if possible, to determine the role of dosage differences in causing disease.

COPY NUMBER VARIATION IN THE HUMAN GENOME

Redon et al. (2006) reviewed CNVs in the human genome. They defined CNVs as variations of segments of DNA 1 kb or larger that are present at variable number in different individuals in comparison with a reference genome. They noted further that CNVs may be simple, involving tandem duplications, or they may be more complex. There is evidence that CNVs influence the phenotype through altered gene expression, for example, in microdeletion and microduplication syndromes. There is evidence that CNVs may play a role in complex diseases. Singleton et al. (2003) reported the occurrence of triplication of the synuclein gene locus in a subgroup of patients with Parkinson disease. Rovelet-Lecrux et al. (2006) reported that duplication of the amyloid precursor protein–encoding locus occurs in a specific form of familial Alzheimer disease.

Copy number variation, involving duplication of specific genes, influences susceptibility to infectious diseases such as HIV-AIDS and to autoimmune disease such as glomerulonephritis (see p. 26).

Redon et al. (2006) noted that analysis of CNV is important since it provides insight into genomic structure. They emphasized the importance of integrating analysis of CNV with data on SNPs. In their study, they analyzed SNPs using Affymetrix SNP gene chips. In addition, they searched for CNV in a whole genome tiling path of large-insert Bac clones. This platform included approximately 26,000 Bac clones that represented 93.7% of the eukaryotic portion of the genome. In total, they identified 1447 CNVs in the human genome. They noted that CNVs often correspond to gaps in the human sequence data. In the Redon et al. (2006) study, the 500K SNP chip was found to be better at detecting small CNVs. The whole genome tiling path of Bacs was found to be better at detecting duplicated genomic regions. They found that deletions are less common than duplications and that the average size of deletions is usually less than 43 kb, while the average duplication size is 120 kb.

Redon et al. (2006) reported that the predominant "functional" sequence elements in the CNV segments are conserved noncoding elements. Genes found enriched in CNV segments include cell adhesion molecules, genes involved in sensory perception of smell, and genes for neurophysiological processes. These investigators reported that CNVs tend to cluster in regions of low coverage in SNP arrays. They noted further that SNP genotypes are significantly impacted by duplications.

Redon et al. (2006) emphasized that regions of CNV represent unstable regions of the genome. Frequently these regions show deletion in some individuals and duplication in others. They reported that a 1 Mb region on chromosome 1q21.1 deleted in patients with cardiac anomalies, was reported to be duplicated in patients with mental retardation and autism spectrum disorder. Inversions also occur in the region.

DELETION POLYMORPHISMS

Conrad et al. (2006) identified 586 distinct regions in the genome that harbor deletion polymorphisms. They reported that normal individuals are hemizygous for 30–50 deletions larger than 5 kb. They discovered deletions through analysis of clusters of SNPs in parent-child trios, followed by searches for apparent Mendelian transmission and SNP genotype errors. These investigators noted that current methods of SNP genotyping would not identify deletion unless trios with both parents and child were analyzed. Conrad et al. followed up on SNP analysis evidence of deletion using comparative genomic hybridization analysis.

A key question to answer is whether deletions are ancient or recent and whether they are recurrent. Conrad et al. (2006) obtained evidence that a number of the deletions they studied represented recurrent deletions since they arose on different haplotypes. A number of the deletion polymorphisms occurred within genes.

Eichler (2006) reported that deletion variants show greater diversity in African populations. Eichler noted that genes that are the site of deletion polymorphisms include environmental interaction genes, such as genes involved in immune defense, drug detoxification, sex hormone metabolism, and cancer susceptibility. Eichler proposed a human genome structural variation project dedicated to characterization of deletions, insertions, and inversion.

GENETIC VARIATION BASED ON REGIONS
OF HOMOZYGOSITY IN THE GENOME

Gibson et al. (2006) analyzed SNP genotypes in different populations and identified genomic regions of many megabases that are homozygous. They defined homozygosity regions as regions of at least 1 Mb in length, with homozygous SNPs. They note that regions of homozygosity for short tandem repeats have been described in the CEPH collections of DNA, representing samples of individuals in different populations distributed worldwide. Homozygous tracts were passed on from parents to offspring. As may be expected, the chromosomal regions where homozygous tracts occur are characterized by low recombination frequencies and high linkage disequilibrium. Gibson et al. reported that of the populations studied, the Yoruba population has the lowest number of homozygous tracts. This is consistent with the hypothesis that the Yoruba population represents an ancient lineage where significant recombination has taken place. In more recent lineages, the time elapsed for recombination to occur is shorter. In some cases a specific region apparently underwent selection in a particular population. A long region of homozygosity occurs around the lactase gene on chromosome 2, which is under positive selection in northern European countries.

It is important to consider whether a region of homozygosity found in an individual represents a transmitted phenomenon common to a specific population or whether homozygosity results from a de novo event in that individual, resulting from deletion or uniparental disomy. Such de novo events may play a role in disease pathogenesis. Deletions may give rise to hemizygosity of a repeated segment that may be reported as homozygosity. In uniparental disomy, both members of a chromosome pair or chromosome segment are derived from one parent, either mother or father.

INVERSION POLYMORPHISMS

Large inversion polymorphisms in the human genome likely arise as a result of nonallelic homologous recombination between low copy repeats with inverse orientation. Inversions are more difficult to detect than dosage changes by cytogenetic analyses and marker analyses. Stefansson et al. (2005) described a 900 kb gene containing inversion polymorphism on 17q21.31. Linkage disequilibrium occurs in this region. The *MAPT* gene (microtubule associated Tau gene) exhibits two widely divergent haplotypes, H1 and H2. The H2 haplotype is uncommon except in Europe. Stefansson et al. demonstrated that this haplotype divergence arose because of an ancient inversion where different sets of alleles were fixed in each configuration. The presence of this inversion polymorphism suppresses recombination locally, resulting in a high frequency of linkage disequilibrium in 17q21.31. The presence of the inversion also leads to copy number changes and altered expression of specific genes in the region.

Detection of 17q21.31 inversion involved analysis of BAC clones containing segments of DNA from this region. They distinguished H1 haplotype and H2 haplotype clones and determined that there were structural differences between the two sets of clones. These structural differences were distinguished by differences in the distance between the genes and differences in gene copy numbers. Stefansson et al. proposed that analysis of patterns of linkage disequilibrium serves as an indirect approach to identifying inversions.

CHROMOSOME INVERSIONS AND GENOMIC IMBALANCE

A number of previous studies have reported occurrence of chromosome inversions in the parents of children with heterozygous deletions, leading, for example, to Angelman syndrome, Williams syndrome, or Soto syndrome. Shaw-Smith et al. (2006) reported the occurrence in three individuals with learning disabilities of deletions encompassing the *MAPT* region on chromosome 17q21.31. In each case, a parent carried the 17q21.31 inversion.

EFFECTS OF STRUCTURAL CHANGES

In order to define the phenotypic relevance of a structural variant, it is necessary to carry out family-based studies to determine if the chromosome change is inherited or de novo. Population data may provide information on the frequency of specific variants. Structural variants may impact gene dosage directly or they may exert position effects. Detailed sequence analysis or SNP analysis may reveal that an apparently balanced translocation is not balanced.

Feuk et al. (2006) proposed that polymorphic CNVs play a role in common complex diseases and in gene–environment interactions. They note, for example, that genes that occur in regions duplicated in CNV often encode proteins involved in metabolism of xenobiotics and response of the organism to external stimuli.

Structural variants may be selectively neutral. They may exert an influence and lead to variation in physiological, biochemical, and pathological processes. In some cases, de novo structural variation may be lethal.

The extent of structural variation in the genome and the size of genomic variants have prompted hypotheses relating to the role of these variants in providing a substrate for evolutionary changes and also in predisposition to common disease (Feuk et al. 2006).

CNVs that contain coding transcripts may alter gene expression and mRNA production. Examples of such CNVs include genes encoding glutathione-S-transferase, cytochrome p450, and complement component 4. CNVs may also affect transcriptional regulation. Goidts et al. (2006) reported that among CNV-associated genes there is an overrepresentation of genes involved in transcriptional regulation. Inversion variants may play a role in disease predisposition directly through alteration of sequence contiguity. Inversions may also predispose to other structural rearrangements. Position effects that influence gene expression include, for example, effects of translocation of a gene into a heterochromatic region. Another epigenetic consideration is whether or not a particular structural rearrangement exhibits a parent of origin effect.

ROLE OF COPY NUMBER VARIATION IN IMMUNE RESPONSE AND DISEASE

Copy Number Variants and Immune-Mediated Glomerulonephritis

Aitman et al. (2006) demonstrated that CNV of the *FCGR3B* gene locus that encodes the receptor for Fc domain of immunoglobulin IgG determines susceptibility to immune-mediated glomerulonephritis. They used quantitative polymerase chain reaction (PCR) to determine *FCGR3B*, the gene copy number in individuals with systemic lupus erythematosis, and lupus nephritis. Their studies revealed that reduced copy number at *FCGR3B* is an independent risk factor for lupus nephritis.

Aitman et al. (2006) noted that Fc receptor-encoding gene loci are located in clusters in mammalian genomes. These receptors link humoral and cellular functions of the immune system and play a role in the immune response. Neutrophils express *FCGR3B*, and this protein tethers neutrophils to immune complexes (antigen-antibody). Aitman et al. proposed that reduced expression of *FCGR3B* may lead to reduced clearance of immune complexes, which in turn may predispose to renal damage.

Copy Number Variants in Chemokine-Encoding Genes and Susceptibility to HIV Infection

Gonzalez et al. (2005) examined a region on chromosome 17q11.2 that includes loci for two chemokine genes, *CCL3L1* and *CL4L1*, in genomic DNA from different individuals. They then carried out studies in different geographic populations to determine if dosage differences in *CCL3L1* influenced the risk of acquiring HIV infection and the progression of HIV disease. Results of their study of transmission in mother-child pairs and of adult-to-adult transmission revealed that low *CCL3L1* copy number led to enhanced HIV susceptibility. They then examined the median population-specific copy number in infected and uninfected groups. Children with less than two copies had a higher chance of HIV infection. Gonzalez et al. demonstrated further that the *CCL3L1* copy number influenced the rate of disease progression. Previous studies demonstrated that the *CCR5* polymorphism influenced risk of acquiring HIV. Studies reported by Gonzalez et al. suggested that low *CCL3L1* dosage provides a permissive genetic background for full expression of *CCR5* genotypes. These findings provide evidence for a direct link between immune response genes and variability in phenotypic response to infectious disease.

It is interesting to note that Bailey et al. (2002) reported that 5% of the human genome contains duplicated sequences enriched for immune response genes.

GENOMIC DISEASES

Lupski first used the term *genomic disease* to define human diseases that arise from microdeletions, microduplications, or structural chromosome rearrangements. Lee and Lupski (2006) noted that genomic rearrangements may give rise to chromosomal disorders, sporadic traits, and disorders that follow a Mendelian pattern of inheritance. They noted further that structural variation and gene dosage changes may contribute to common complex disorders. Within the category of genomic rearrangements, Lee and Lupski included duplications, deletions, inversions of specific genomic segments, and more complex rearrangements such as translocations, marker chromosomes, and isochromosomes. They reviewed evidence that region-specific repeat sequences increase the likelihood of structural rearrangements. Recurrent rearrangements are characterized by their occurrence in unrelated individuals, and these often have common breakpoints. Recurrent rearrangements result from non-allelic homologous recombination. The mechanisms that predispose to nonrecurrent rearrangements and nonhomologous end joining are still not clearly defined.

VARIATION IN IMPACT OF GENOMIC DOSAGE CHANGES BASED ON GENE EXPRESSION

Prandini et al. (2007) studied patients with Down syndrome to investigate their hypothesis that phenotype variation among trisomy 21 Down syndrome patients was due to individual differences among the patients in the degree of overexpression of gene on chromosome 21. They used mRNA derived from fibroblasts and lymphoblastoid cell lines and microarrays to quantify differences in gene expression levels. They reported that for 39% of genes in lymphoblastoid cell lines and 62% of genes in fibroblasts, the steady-state RNA levels of chromosome 21 genes in these samples was on average 1.5, as would be expected for trisomy. They classified genes into three types: genes that were overexpressed; genes that showed a variable degree of overexpression and individual differences in degree of gene expression; and genes that were dosage insensitive and showed the same degree of expression in patients as controls. These studies indicate that it is important to consider natural variation in gene expression in aneuploidies. Genes that show variable degrees of expression in the trisomy 21 patients may account for interindividual differences in these patients.

RESOURCES TO SEARCH FOR GENOMIC DISEASES DUE TO DOSAGE CHANGES

Cytogenetics and Fluorescence in Situ Hybridization

Classical karyotyping used in clinical cytogenetics involves microscopic analyses of banded chromosomes. These studies can detect deletions or duplications larger than approximately 5 Mb. The combination of microscopy and the use of labeled DNA probes for chromosome analysis in fluorescence in situ hybridization (FISH) studies improve resolution so that deletions or duplications of between 1 and 3 Mb can be detected.

The availability of human DNA sequence information has enabled development of new technologies, particularly microarray technologies, for assessment of genomic composition and imbalance.

Bac Arrays

Array comparative genomic hybridization is often carried out using arrayed Bac clones with large inserts of human genomic DNA. The average insert clones in these arrays is approximately 170 kb. Considerable refinements in the selection of the Bac clones to be used on arrays have taken place, to decrease the signal-to-noise ratio. Genomic DNA derived from test and control samples is differentially labeled. For example, in one hybridization experiment test DNA is labeled with a green fluorescent dye and control DNA is labeled with a red fluorescent dye. In a parallel experiment, the test DNA is labeled with a red dye and control DNA is labeled green. The labeled genomic DNA and repeat blocking DNA are hybridized to the microarrays in individual experiments. Following hybridization and washing, microarrays are scanned and the ratio of the signal with the red versus the green dye is then measured.

Oligonucleotide Arrays

These consist of short DNA probes, often 35–50 nucleotides in length, that correspond to unique sequence DNA attached to a solid matrix.

SNP Arrays

In SNP arrays, nucleotide sequences corresponding to mapped segments of DNA that contain SNPs are attached to a solid matrix. It is estimated that 10 million SNPs occur in the human genome, one SNP every 300 nucleotides. The signal-to-noise ratio generated in scans of microarrays has steadily improved through advances in photolithography processes that synthesize oligonucleotides directly on a solid matrix, modifications to array surface chemistry, and improved scanning technology

The application of extended SNP data and of microarray data to analysis of human genetic disease requires that background information be gathered to determine the extent of normal variation in the genome in the population from which cases are drawn. Fortunately this information is available in accessible databases.

Matsuzaki et al. (2004) reported advances in SNP detection through array technology that permits genotyping of over 100,000 SNPs. Newer technologies make use of the new data on the SNP content of the genome. In addition, they report that higher signal-to-noise ratios result from advances in photolithography processes that bind oligonucleotides to solid matrix, modifications to array surface chemistry, and improved scanning technology. Protocols for use of the 100 k Affymetrix SNP chip include digestion of genomic DNA in parallel with *Xba*1 and *Hin*dIII restriction endonucleases. The restriction fragments are then ligated to adaptors and amplified in PCR with Pfx polymerase that yields fragments in the 250–2000 bp size range.

The microarray platform Affymetrix Human SNP 5.0 Gene Chip provides 50 times greater resolution in searching for genomic variation than that provided in the previous SNP arrays. These new arrays are designed so that eight different probes represent each SNP. Hybridization intensities will then vary according to homozygote and heterozygote status. The SNPs selected to produce these arrays have unique map assignments on human chromosomes. On these arrays, the mean intermarker distance is 2.5 kb.

In addition to the 500,568 SNPs on these arrays, there are an additional 420,000 nonpolymorphic probes that yield data to determine copy number changes.

Defining Dosage by Quantitative PCR

Pathologically significant copy number changes may involve whole chromosomes, segments of chromosomes, or exons. PCR amplification will not readily detect heterozygous deletions unless it is designed to be quantitative. Schouten et al. (2002) designed multiplex ligation-dependent probe amplification (MLPA). Probes are specifically designed to straddle the site of interest in the genome (the target). Probes contain a unique sequence corresponding to the target sequence and in addition they contain a sequence common to each probe a universal, primer sequence segment. Probes are ligated to the DNA sample to be tested. Following ligation, probes are amplified using the universal primer sequence. MLPA is useful for analysis of ploidy, for dosage, or for analysis of specific genes.

Assessment of Chromosomal Translocations and Breakpoint Analysis

Balanced translocations may lead to phenotypic changes (genomic disease) through gene disruption, or separation of a gene from its *cis*-regulatory elements. In addition, translocations that appear microscopically balanced may in fact, at a molecular level, be associated with submicroscopic deletions or duplications. In some cases, however, the phenotype found in a patient may not be the direct result of the translocation.

Fluorescence in situ hybridization studies on metaphase chromosomes or on interphase nuclei may sometimes lead to the identification of Bac clones that are split as a result of a balanced translocation. Exact identification of the chromosome breakpoint is, however, not directly possible from FISH experiments.

Breakpoint analysis is facilitated by separation of the derivative chromosomes from the normal homologue. Such separation may be achieved by use of somatic cell hybrids or by flow sorting of chromosomes. Following separation of the translocation products, the derivative chromosomes are differentially labeled, for example, with either red or green fluorescent dye. The translocated chromosome material is therefore labeled differently than the material in its normal homologue. The labeled material can then be used in array analysis (Fiegler et al. 2003).

Gribble et al. (2007) reported the use of a three-tiered microarray approach to translocation breakpoint identification. The first step in the analysis involves hybridization to microarrays of large-insert Bac clones. In these clones, the average size of human genomic inserts is approximately 170 kb. Analysis of the ratio of hybridization of the differentially labeled sorted translocation chromosomes to the microarrays provides low-resolution information on the position of the chromosome breakpoint. Tiling path clones with smaller inserts of human genomic DNA, average size 40 kb, in cosmids or fosmids are then used in hybridization experiments with sorted labeled translocation chromosomes. These experiments achieve finer localization of the breakpoint.

Oligonucleotide probes are generated from DNA present in the region of the breakpoint identified in tiling path clones. These oligonucleotide probes, 45–77 bp in length, are used to construct microarrays for ultra-high-resolution analysis. Information from this analysis is then used to design primers for PCR amplification of translocation junction fragments and sequencing.

CLINICAL APPLICATIONS OF HIGH-RESOLUTION GENOMIC ANALYSIS

Use of Oligonucleotide SNP Arrays to Detect Chromosomal Imbalances

Ming et al. (2006) reported on the use of Affymetrix SNP arrays to detect known chromosomal dosage changes in proof of principle experiments. Subsequently they used these arrays to detect previously unrecognized deletions and rearrangements in patients with chromosome abnormalities. In each case, the abnormality detected in the child was found not to be present in the parent. If dosage differences found in a child were due to copy number polymorphisms, the parents would show the same variation as the child. It is important to note that the density of SNP probes available in a particular chromosome region will affect the sensitivity of the analysis. Ming et al. noted that statistical significance of the copy number changes was usually reached when five probes in a row were deleted.

Determining the Significance of Structural Chromosome Changes

Usually, structural chromosome abnormalities in patients with a specific disorder are considered causative if they are present in an affected offspring but not in the unaffected parent. However, it is important to consider the sex of the transmitting parent. It is well known, for example, that duplications in the chromosome 15q12-q13 region are more likely to be associated with phenotypic abnormalities if they are present on a maternally derived chromosome. The sex of the affected probands must also be taken into account. Parent-of-origin effects also appear to play a role in duplications of chromosome 16p13 (Ullmann et al. 2007). Duplications and deletions of this chromosome segment arise, likely because of the genomic architecture of the region and associated long repetitive elements. There are reports of 16p13 imbalances in subgroups of patients with autism and subgroups of patients with mental retardation.

Diagnostic Genome Profiling in Mental Retardation

De Vries et al. (2005) carried out array-based copy number analysis in 100 patients with unexplained mental retardation. Prior conventional chromosome studies and analyses to detect subtelomeric chromosome changes yielded normal results. De Vries et al. used a 32,447 clone tiling resolution genome array. They found clinically relevant de novo alterations in 10 patients. The alterations varied in size from 540 kb to 12 Mb.

Analysis of Sequences in Rearrangement Hotspots

Sharp et al. (2006) identified 130 sites that are hotspots for chromosome rearrangements, based on the presence of interspersed segmental duplications greater than 1 kb in length that show .90% sequence homology. They developed a Bac array that specifically targets these hotspot regions. Inserts in the Bac clones comprise nonredundant DNA sequences located between the duplicated sequences. Using this array, they analyzed DNA from 290 individuals with mental retardation who were reported to have normal karyotypes at the 550 band level. They identified pathogenic structural chromosome changes in 16 patients. These changes included deletions, duplications, and rearrangement and were located on 1q21.1, 15q13, 15q23-24, and 17q12.

Genomic Profiling in Autism

Autism spectrum disorders (ASD) are neurodevelopmental disorders. The diagnostic features include deficits in reciprocal social interactions and communications, and restricted patterns of behavior and interests. There is definite evidence that genetic factors play a role in autism. Twin studies revealed that monozygotic twins show 60% to 90% concordance for autism while concordance in dizygotic twins varies between 0% and 10%. ASD occurs in 5% to 10% of siblings of ASD probands. Studies to date indicate genetic heterogeneity in autism.

Routine cytogenetic studies reveal microscopic chromosome abnormalities in 10% to 15% of cases of autism (Martin and Ledbetter 2007). Autism may arise in a number of different chromosome abnormalities.

Two recent studies have revealed, through microarray analyses, that genome dosage changes are common in autism. Analyses reported by the Szatmari et al. (2007) were restricted to families with two or more autism-affected individuals. The microarray platform used was the 10k Affymetrix SNP array. The Autism Genome Project study revealed that genome CNVs occurred in 7.6% to 11% of affected individuals, depending on the type of analysis used (size of sample batch). This study led to identification of a deletion in chromosome 2p16 in one pair of siblings with autism. This deletion eliminated the neurexin 1 gene *NRXN1*. Of particular interest is the fact that neurexin interacts with neuroligins at neuronal synapses. Previous studies demonstrated deletions or mutations in the neuroligin genes *NGLN3* and *NGLN4* in small subgroups of autism subjects (Jamain et al. 2003).

In a study reported by Sebat et al. (2007), de novo genome CNVs (variations occurring in autism-affected offspring but not in parents) were significantly associated with autism, $p = 0.0005$. CNVs were more strongly associated with sporadic autism (10%) of cases than with familial autism (2%). In the study reported by Sebat et al., CNV occurred in 1% of controls. Methods used to detect CNV included oligonucleotide microarray analyses, microscopy with FISH, and genotyping with microsatellite polymorphisms. Sebat et al. (2007) concluded that de novo germline mutation is a more significant risk factor for autism than previously recognized.

HIGH-RESOLUTION CYTOGENETIC ANALYSES FOLLOW-UP AND GENE ANALYSIS: IDENTIFYING KEY GENES THAT LEAD TO PHENOTYPE

The location and extent of chromosome dosage changes can be accurately determined by microarray technologies. Bac arrays, containing large segments of human DNA cloned into Bac vectors, and oligonucleotide arrays are particularly useful for dosage studies. SNP arrays to search for polymorphism may be useful in dosage studies and in addition may provide information on structural chromosome changes including translocations and inversions. High-resolution cytogenetic analysis is yielding information on the etiology of unexplained mental retardation. In such cases, it is important to carry out studies on parents. A parent with a normal clinical phenotype may carry a structural chromosome abnormality, such as a translocation or inversion that predisposes to chromosome imbalance in the offspring. Furthermore, it is possible that gonadal mosaicism exists in a parent and that two cell types are present in the gonad, one with normal chromosome complement and another with altered chromosome dosage. Families should be appropriately counseled regarding options for prenatal diagnosis in subsequent pregnancies.

Another important aspect of high-resolution genomic analysis of clinical syndromes is the finding that in a number of syndromes, deletions of one particular gene are largely responsible for the features of the syndrome. Furthermore, intragenic deletions and certain mutations involving these genes may give rise to the characteristic features (Feenstra et al. 2006). Examples of syndromes where large genomic deletions or deletions or mutations of a single gene lead to typical syndrome features include Smith Magenis syndrome, Sotos syndrome, 9q34 deletion syndrome, and 22q13 deletion syndrome.

Smith Magenis Syndrome and Retinoic Acid–Induced Gene I

Approximately 75% of patients with Smith Magenis syndrome have deletions that span 3.5 Mb on chromosome 17p11.2. Features of this syndrome include distinct craniofacial features including midfacial hypoplasia, deep-set closely spaced eyes, prominent chin, dental anomalies, and middle ear and laryngeal anomalies. Characteristic neurobehavioral anomalies include developmental delay and self-injurious and stereotypic behaviors. Several investigators have described patients with typical features of this syndrome who have no evidence of deletion on chromosome 17p11.2 and have determined that intragenic deletions or mutations in retinoic acid–induced gene 1 (*RAI1*) may give rise to the syndrome. *RAI1* maps in the midpoint of the large deletion region.

22q13 Deletion Syndrome and *SHANK3*

Features of the 22q13 deletion syndrome include neonatal hypotonia, severe expressive language delay, developmental delay, and autistic behaviors. Most commonly, deletions of between 3.3 and 8.4 Mb give rise to this syndrome. However, the syndrome may arise through translocations. These disrupt a specific gene, the *SHANK3* (*ProSap2*). This gene encodes a protein that plays a role in the structural organization of dendrites. Mutations in *SHANK3* occur in rare cases of autism. *SHANK3* interacts with neuroligins, which may also be mutated in some forms of autism (Durand et al. 2007).

9q34 Terminal Deletion Syndrome and Euchromatic Histone Methyl Transferase Enzyme *EHMT1*

The 9q34 terminal deletion syndrome is characterized by severe mental retardation, hypotonia, microcephaly, and typical facial dysmorphology characterized by midfacial hypoplasia, prominent chin, and everted lower lip. The features of this syndrome have also been found in patients with translocations that disrupt a specific gene, *EHMT1*, that encodes a euchromatic histone methyl transferase enzyme that specifically reacts with histone H3 in core histones (Kleefstra, Brunner, et al. 2006; Kleefstra, Koolen, et al. 2006).

Collectively, these cases illustrate the point that it may be necessary to examine specific single genes in patients with features of microdeletion syndromes who do not show the typical genomic deletions.

3

SIGNIFICANCE OF DNA SEQUENCE
CHANGES FOR DIAGNOSIS AND THERAPY

As information on genomic sequences increases and opens the way for analyses to determine the role of DNA sequence changes in determining phenotypic variation and disease, it becomes increasingly important to revisit and expand concepts regarding the functional significance of specific sequence changes.

DNA sequence changes include nucleotide substitutions, deletions, and insertions. Simple mutations involve a single nucleotide. Copy errors during DNA replication may lead to mutation; such copy errors are, however, minimized by the proofreading capacity of DNA polymerase. More complex changes involve the exchanges between alleles at homologous loci or exchanges between nonallelic sequences (Strachan and Read 2004).

DNA polymorphisms are defined as allelic variants that occur with a frequency of 1 in 100 (0.01) individuals in the population. Analysis of the human DNA sequence revealed that the average diversity (average heterozygosity) is 1 in 1250 nucleotides. The degree of heterozygosity differs in different regions of the genome.

Mutations or transitions that involve the substitutions of pyrimidine for pyrimidine, cytosine to thymine (C to T), or purine for purine, adenine to guanine (A to G), are more common. Mutations may involve coding sequences. Exon sequences are more highly conserved between individuals (85% identity) than intron sequences (69.1% identity). The degree of conservation of untranslated sequences at the 5' and 3' ends of genes and promoter sequences is approximately 75%. The substitution rate varies on different chromosomes and in different chromosome regions (Strachan and Read 2004). In synonymous mutations in coding regions the nucleotide change does not lead to an amino acid change. This is based on the fact that the code is degenerate and for many amino acids, a number of different nucleotides may encode the same amino acid. In nonsynonymous mutations, the code is changed, and this may lead to amino acid substitution. Such missense mutations may be defined as conservative if the amino acid is replaced by a similar amino acid. Grantham (1974) determined the degree of similarity of different amino acids and generated a table that defines degrees of relationship. This is discussed further below (on p. 47).

Nonsynonymous substitutions may lead to generation of a stop codon, resulting in abnormally short messenger RNA (mRNA) transcripts. Some of these may be destroyed through nonsense-mediated decay. Deletions or insertions that result in a codon frame shift result in synthesis of abnormal mRNA and may lead to production of unusual proteins.

VARIATION IN REPEAT SEQUENCES

A common form of DNA polymorphism involves variation in the number of tandem repeats. These repeats include simple repeats, such as dinucleotide or trinucleotide repeats (microsatellites) or more complex repeats. Polymorphism in microsatellite repeats may arise because of staggered pairing of the two complementary strands. Transposon repeats, such as Alu, LINE, and LTR repeats, often vary in length and contribute to polymorphism.

More complex mutations may arise because of recombination between nonallelic but similar DNA sequences that are present in tandem repeats. Gene conversion involves the nonreciprocal transfer of a DNA segment. This transfer may occur between alleles or between loci. Large-scale structural polymorphism found in the genome includes segmental duplications, deletions, and inversions (see Chapter 2). DNA may also undergo changes through interactions with chemicals, metabolites, and ionizing radiation (see Chapter 8).

DEFINITION OF A GENE, GENETIC CODES, AND TRANSCRIPTION CIRCA 2007

Consensus Description of a Gene

Carninci (2006) noted that the term *gene* was initially defined as a unit that controls a trait. Subsequently it was defined as a unit of DNA that produces mRNA and encodes a protein. He concluded that new nomenclature might be required to describe different elements of the transcriptome.

The following consensus description of a gene was presented by Pearson in 2006: "a locatable region of genomic sequence corresponding to a unit of inheritance, which is associated with regulatory region, transcribed region and/or other functional sequence regions" (p. 401).

RNA Code

The first RNA code is the three-letter nucleotide code that determines the specificity of amino acids to be added to the peptide chain that results from mRNA translation. More recently, molecular biologists have defined a "second code" that determines maturation and processing of mRNA prior to translation (Guigo and Valceral 2006). mRNA transcript processing includes removal of exons and splicing of introns. Differences in processing of a specific mRNA transcript give rise to alternate processed transcripts. Guigo and Valceral noted that splice site sequences are degenerate and intron-exon boundaries are often difficult to distinguish. More recent studies have also revealed that sequences deep within exons and introns may impact splicing. Understanding of the splicing code is important in determining effects of mutations and in developing concepts related to tissue-specific gene expression.

There is increasing evidence for the existence of overlapping or fused gene transcripts. Fused transcripts start at the 5′ end of one gene and run through to the 3′ end of the downstream gene. Sequence between the two genes is spliced out as an intron. Akiva et al. (2006) carried out systemic studies to identify fusion transcripts in humans. These studies included examination of cDNA and expressed sequence tag (EST) databases and subsequent experimental validation of proposed fused genes by reverse transcriptase polymerase chain reaction. They reported that the average distance between genes that fuse is 48 kb; however, in 5% of fused genes the contributory genes were more than 50 kb apart. These investigators discovered unusual fusion genes that resulted from fusion of the 5′ untranslated region or first exon of one gene with the downstream gene, resulting in a gene with the regulatory characteristics of the upstream gene. Akiva et al. (2006) noted that in some cases translation of the fused gene transcript resulted in the generation of a bifunctional protein. Their studies led them to conclude that chimeric transcripts occur more frequently in the human genome than previously recognized and that this fusion process leads to additional protein variation and to changes in gene regulation.

Transcription of DNA That Does Not Encode Proteins

There is an increasing body of evidence that RNA transcripts are generated not only from protein-coding regions of the genome but also from noncoding regions. RNA transcripts derived from the latter regions may be polyadenylated and may be transported from the nucleus to the cytoplasm (Cheng et al. 2005). However, the noncoding RNA transcripts make up a large proportion of the unpolyadenylated transcriptome. Many of the noncoding RNAs do not leave the nucleus (Carninci 2006). The exact roles of these transcripts have not yet been defined.

Noncoding RNAs play a role in alternate splicing. For example, the small nucleolar RNA, snoRNA HBII52, encoded in the Prader-Willi region of chromosome 15 (15q12), regulates alternate splicing of the X-linked serotonin receptor 21 transcript. Prader-Willi syndrome patients lack the HBII52 transcript (Kishore and Stamm 2006).

Noncoding RNAs play a role in the regulation of expression of imprinted genes and have a function even though they do not leave the nucleus. There is growing evidence that many genes give rise to both sense (S) and antisense transcripts (AS). S/AS pairs are often differentially expressed in different tissues and under different metabolic conditions; furthermore, S/AS transcripts may regulate expression of each other (Carninci 2006).

MUTATIONS THAT IMPACT mRNA PROCESSING

Transcript Processing

Mutations in noncoding regions may alter gene expression through their effects on mRNA transcription or mRNA processing. This includes splicing, capping, and polyadenylation. Capping involves the linkage of 7 methyl guanosine to the 5' carbon of the first nucleotide. Specific sequences at the 3' end of genes act as polyadenylation signal sequences.

At the 5′ end of an intron, in the splice donor site, GT dinucleotides occur (GU in mRNA) and AG dinucleotides occur at the 3′ end of the intron, at the splice acceptor site. Other sequences important for splicing are the poly-pyrimidine tract that precedes the end of the intron and splice enhancer and splice silencer sequences. A conserved sequence at the branch site within an intron is important in lariat formation during splicing out of introns. Small ribonucleoprotein particles, snRNPs, bind to the splice donor site, the splice acceptor site, and to the branch site.

Alternate Splicing

Sequencing of genomes of different organisms has revealed that mammalian species and the plant *Arabidopsis thaliana* have similar numbers of genes, 25,000–30,000. Furthermore, mammalian genomes have only four times as many genes as yeast. It has also become clear that many protein-coding segments can be transcribed in different ways to generate different protein variants and that these may have different functions. There is evidence that the same segment of DNA may be transcribed in opposite directions. Full understanding of the role of DNA sequence changes in generating macromolecular and cellular complexity and in determining phenotype requires that alternate transcription be analyzed (Blencowe 2006).

Evidence of the high frequency of alternative splicing has been gained through analysis of genomic sequence data and comparison with mRNA, cDNA, and short expressed sequences, ESTs, by in silico means or through use of microarrays. At least 59% of human genes undergo alternate splicing and 80% of alternate transcripts result in protein coding changes (Faustino and Cooper 2003). There is evidence that alternative splicing may be cell or tissue specific (Slaughenhaupt et al. 2001). Cell-specific differences in splicing may play a role in the cell- and tissue-specific effects of different mutations.

High Throughput Analysis of Splicing

High-throughput experimental approaches are being adopted to analyze alternate splicing on a genome-wide level. Genomic sequence analysis programs may be used to search for specific sequence elements that are important in splicing. In addition to databases that provide sequences of cDNAs and ESTs, specific databases have been developed to compile alternate transcripts.

In these microarrays designed to analyze alternate splicing, thousands of oligonucleotide probes are bound to a solid matrix or to fibers. Oligonucleotide probes may be primarily selected that correspond to splice junctions

(Castle et al. 2003). Oligonucleotide probes corresponding to sequence elements that play a role in tissue-specific alternate splicing were also included in the microarrays developed by Sugnet et al. (2006).

Different Forms of Alternate Splicing

The most common form of alternate splicing involves inclusion of alternate exons (Blencowe 2006). Based on selection of alternate $5'$ or $3'$ splice sites, a particular exon may be skipped or included. The resulting spliced transcripts may then differ by one or more exons. Blencowe noted that in some cases where there is a mutually exclusive exon inclusion, spliced transcripts contain only one or the other exon. For example, in alpha tropomyosin, transcripts contain only exon 2 or exon 3.

Alternate transcripts may result from intron inclusion. Transcripts of different length may result from the choice of alternate $5'$ transcription start sites or different $3'$ transcription termination sites. Splice site selection is based on the binding of regulatory factors to specific sequence elements.

Cellular differentiation is often associated with the presence of alternate transcripts (Watson et al. 2005). Ule and Darnell (2006) reported that alternate splicing is linked to differential neuronal activity and synaptic plasticity. The presence or absence of exon 20 affects trafficking of N-methyl-D-aspartic acid neurotransmitter receptor (NMDAR) from the endoplasmic reticulum to the synapse. Alternate splicing of exon 19 regulates NMDAR localization and phosphorylation. Ule and Darnell noted that RNA binding proteins impact splicing.

An important consideration is whether or not specific alternate transcripts are functionally active. Alternate transcription may introduce termination codons. Such transcripts are often destroyed in the process of nonsense-mediated decay. Noncoding RNA transcribed from introns or intergenic regions may have a physiological role.

Through alternate splicing, individual genes can express multiple mRNAs. These different mRNAs may encode proteins with different functions. In some cases, the different transcripts may encode proteins with alternate functions. Blencowe (2006) noted that there was little information on the coordination or regulation of splicing on a broader scale. The question arises as to whether low-abundance splice variants have a physiological effect, and whether the generation of alternate transcripts results in evolutionary advantage.

Cis-acting splice mutations affect the splicing of specific DNA segments. *Trans*-acting splice mutations may affect splicing of a number of different genes or DNA segments (Faustino and Cooper 2003). *Trans*-acting mutations may arise through mutation in protein components of the splicing machinery.

Soret et al. (2005) reported that mutations in serine-arginine-rich (SR) proteins that bind to exonic splice enhancers (ESE) may impact ribonucleoprotein binding and ESE action. There is also evidence that transcription of specific genes may be impaired by trinucleotide repeat expansion or the presence of polyglutamine repeats that bind transcription factors.

This is apparently the case in myotonic dystrophy (DM1), where expansion of trinucleotide repeats at the 3′ end of the gene leads to sequestration of muscle tissue–specific splicing factors. Miller et al. (2000) proposed that DM1 disease is caused by aberrant recruitment of the muscleblind proteins (EXP) to the DMPK transcript (CUGn) expansion. EXP proteins are homologues of the muscleblind proteins first described in *Drosophila* and known to play a role in the terminal differentiation of muscle.

Jiang et al. (2006) reported that mutant huntingtin, through its binding to CREB-binding protein, depletes this transcription factor and impairs transcription of a number of genes.

Splice Enhancers and Silencers and Splice Regulatory Sequences

Correct splicing is not only dependent on sequences at exon-intron boundaries. Exonic enhancer and silencer sequences and intronic enhancer and silencer sequences interact with proteins to regulate splicing. Nielsen et al. (2007) noted that exonic mutations at such sites may have different effects than those predicted on the basis of their effects on protein coding.

There is evidence that in some instances, synonymous nucleotide variations may alter splicing. These effects will not be evident unless mRNA studies are carried out. Nielsen et al. (2007) noted that search algorithms for splice enhancer and splice silencer sequences are under development but may still be imprecise; functional characterization of the effects of sequence changes are required. They studied the effect of two different point mutations in exon 5 of the gene that encodes medium-chain acyl-CoA dehydrogenase (MCAD). They demonstrated that a synonymous mutation impacted splicing and noted that the DNA sequence in the vicinity of this synonymous mutation is highly similar to sequence in the *SMN 2* exon 7 gene that is associated with abnormal splicing. An exonic mutation in MCAD, c.362 C-T, led to skipping of exon 5 and to a significant decrease in MCAD mRNA in the patient's cells. They noted that this mutation likely disrupts a splice enhancer because it alters the affinity of this splice enhancer for a splice factor SF2/ASF. This is a serine-arginine-rich splicing factor also known as SFRS1. The exon 5–depleted mRNA is apparently degraded by nonsense-mediated mRNA decay.

Nielsen et al. (2007) further demonstrated that an exon 5 polymorphism that leads to a synonymous coding change c.351 A-C neutralized the negative effect of the c.362 C-T mutation. They concluded that the potential effects of mutations should be evaluated in the context of the haplotype.

ALTERNATE SPLICING AND SPLICE REGULATION RELEVANCE TO DIAGNOSIS

New information on alternate splicing, splice regulation, and functions of splice enhancers and silencers has relevance for diagnosis of genetic diseases. Bergmann et al. (2006) reported on difficulties in determining the functional significance of DNA sequence changes in the gene *PKHD1*. Mutations in this gene lead to the autosomal recessive form of polycystic kidney disease, ARPKD. This gene maps to chromosome 6p12. The longest open reading frame of *PKHD1* contains 66 exons. Various alternate transcripts occur due to a complex pattern of splicing. Mutations are found in approximately 80% of cases with ARPKD. Bergmann et al. (2006) suggested that changes currently classified as nonpathogenic may indeed be pathogenic. They noted that variants residing outside coding regions or in regulatory regions may be missed. The functional significance of nucleotide changes in splice regulatory elements in introns may be difficult to assess. Bergmann et al. identified a splice site mutation in an infant diagnosed with ARPKD. They demonstrated that mutation in intron 2 leads to skipping of exon 3. They describe the construction of a minigene and transfection studies to investigate the functional effects of the intron mutation.

Chen et al. (2006) analyzed intronic sequence alterations in the *BRCA1* and *BRCA2* genes to determine pathogenicity. They undertook analysis of mRNA transcripts in patients with germline mutations at or near intron splice junction sequences. In seven cases they determined that there was no apparent effect on mRNA splicing. In four cases the intron sequence changes caused frameshift mutations and resulted in deletions.

SPLICE MUTATIONS IN HUMAN DISEASE

Frequency in Human Genetic Diseases

Faustino and Cooper (2003) estimated that splicing defects represent 15% of disease-causing mutations. Splicing defect mutations frequently produce both wild-type and mutant forms of mRNA, and the severity of the phenotype is related to the proportion of the two forms (Tang et al. 2006). Splice mutations that lead to complete absence of a specific exon may be associated with severe disease phenotype. A less severe disorder results when mutants are leaky and produce a low percentage of transcripts. Mutations in the *ATP7A* gene serve as an example. *ATP7A* encodes ATPase, Cu^{++} transporting, alpha polypeptide.

Menkes Disease, Occipital Horn Syndrome, and *ATP7A* Deficiency

ATP7A is required for intracellular transport of copper to copper-requiring enzymes such as lysyl oxidase and dopamine beta-hydroxylase. The enzyme is encoded by a gene on the X (Xq12-q13) chromosome and is deficient in both Menkes disease and occipital horn syndrome. Menkes disease is characterized by neurological degeneration, connective tissue abnormalities (manifestations include hernias, diverticulosis, and aneurysms), and abnormalities of the hair and skin. In occipital horn syndrome the clinical manifestations are milder. The occipital horns are bony prominences on either side of the foramen magnum.

ATP7A deficiency may be due to chromosome abnormalities, gene deletions, or mutations. At least 30% of the mutations in this disorder are splice site mutations. Splice site mutations are found more commonly in occipital horn syndrome than in Menkes disease (Moller et al. 2000). Moller et al. reported that the more severe disorder, Menkes disease, and the milder disorder, occipital horn syndrome, occurred in different patients with mutations at the same splice site and led to absence of exon 6 in transcripts. They determined that in patients with the milder phenotype, correctly spliced exon 6 transcripts occurred at a level that was 2% to 5% of that in unaffected individuals.

SPLICE ALTERATIONS AND MUTATIONS: THERAPEUTIC POSSIBILITIES

Spinal Muscular Atrophy

The disorder spinal muscular atrophy (SMA) is associated with progressive loss of alpha motor neurons in the anterior horns of the spinal cord. This leads to weakness and atrophy of voluntary muscles. Initially the proximal limb muscles are involved. Nerve conduction velocities are significantly decreased.

There is no involvement of sensory nerves. SMA is classified on the basis of the age of onset. The most severe form, Werdnig-Hoffman disease, has onset during the first 6 months of life. The intermediate or juvenile forms have onset during childhood. The mildest form has onset during adult life and does not affect life expectancy. Wirth et al. (2006) reviewed the genetics of SMA and possible therapeutic interventions. SMA is due to a defect in the *SMN1* gene that maps to chromosome 5q13. Duplicate copies of this gene occur in this chromosomal region. Differences between *SMN1* and *SMN2* occur at the $3'$ end of the gene, and a C-T transition occurs in exon 7 of *SMN2*. This change affects splicing so that exon 7 is not present in 90% of the *SMN2* transcripts. Furthermore, the quantity of *SMN2* mRNA is greatly reduced. An exonic splice enhancer (ESE) is present in exon 7 of *SMN1*. This ESE is destroyed by the nucleotide substitution in *SMN2*, resulting in decreased recruitment of the splice-acting proteins. Absence of the ESE and bound protein leads to exon skipping. The *SMN2* protein is 282 amino acids in length while the *SMN1*-encoded protein is 294 amino acids in length.

Wirth et al. (2006) reported that 96% of patients with SMA types I, II, and III have homozygous deletions of *SMN1* exons 7 and 8 or 7 only. Heterozygous deletions occur in 4% of patients and these patients have mutations in the *SMN1* gene on the other chromosome. In different individuals, the *SMN2* copy number differs and from one to eight copies may be present. Higher *SMN2* copy numbers occur in patients with the mildest form of SMA, type IV. In a patient with absence of *SMN1*, the phenotype can be predicted by determining the *SMN2* copy number.

Therapies in SMA include use of neurotrophic factors and exercise to improve motor neuron viability. In addition, therapies in SMA are designed to modulate splicing and transcription of *SMN2*. Histone deacetylase inhibitors, such as valproic acid and 4-phenylbutyrate, have been shown to elevate *SMN2* transcription and generation of transcripts in which exon 7 is included (Brichta et al. 2003). Valproic acid treatment resulted in clinical improvement in 50% of patients. These investigators reported that the upregulation of *SMN2* in response to valproate was most likely due to increased expression of the SR-like (serine-arginine like) splicing factor Htra2-beta 1.

Weihl et al. (2006) reported results of valproate treatment of seven adult patients with SMA type III/IV over a period of 8 months. Treatment resulted in increased quantitative muscle strength and improved subjective function.

In SMA, information on the molecular pathogenesis and knowledge of splicing mechanisms and transcription control therefore have direct application to therapeutic intervention.

Familial Dysautonomia

Familial dysautonomia is a recessively inherited neurodegenerative disease characterized by autonomic dysfunction and sensory neuropathies. The carrier frequency for familial dysautonomia in Ashkenazi Jews of Polish descent is 1 in 18. Anderson et al. (2001) and Slaugenhaupt et al. (2001) discovered that this disorder is caused by mutations in the *IKBKAP* gene (inhibitor of kappa light polypeptide gene enhancer in B-cells, kinase complex-associated protein) that encodes a subunit of the Elongator complex that plays a role in transcription elongation. *IKBKAP* maps to chromosome 9q31. Among patients, 99% are homozygous for an intronic mutation that leads to variable skipping of exon 20. These patients produce a low percentage of full-length IKAP protein. Furthermore, there is evidence for tissue-specific differences in the relative quantity of full-length and exon 20-deleted transcripts (Slaugenhaupt et al. 2001). Splicing defect mutations frequently produce both wild-type and mutant forms of mRNA, and severity of the phenotype is related to the proportion of the two forms (Hims, el Ibrahim, et al. 2007; Hims, Shetty, et al. 2007; Figure 3–1).

Slaugenhaupt et al. (2004) reported that the plant cytokinin known as kinetin rescues mRNA splicing in this disorder. Hims and colleagues (Hims, el Ibrahim, et al. 2007; Hims, Shetty, et al. 2007) used a minigene transfection model to investigate the molecular mechanism of the kinetin effect. They identified a sequence CAA at the 3′ end of exon 20 that is essential for this effect. It is possible, however, that other sequences may influence the effect. They noted that kinetin affects splicing in a number of other genes, and they showed correction of a splice defect in NF1 in neurofibromatosis through use of kinetin.

Other Small Molecules That Target Splice Regulators

Splice site recognition is dependent on the interaction of specific exonic and intronic sequences in the pre-mRNA with two classes of nuclear RNA binding proteins: SR proteins and heterogeneous nuclear ribonucleoproteins. SR proteins bind to exonic splice enhancers. This binding promotes exon definition and prevents activity of new splice enhancers. Soret et al. (2005) proposed that small molecules that target splice regulators may be used to correct splicing. They noted that indole compounds inhibited a subset of SR-binding proteins.

Translational Aspects

There is evidence that therapeutic agents including kinetin and indoles may impact splicing and correct certain disease specific splicing mutations. Anti-

A. Exons 19, 20, 21 and introns, wild-type *IBKAP* Pre-mRNA

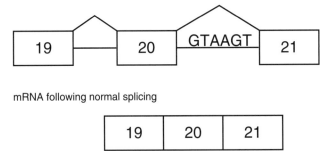

mRNA following normal splicing

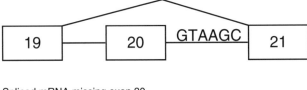

B. Exons 19, 20, 21 and introns, intron mutation leads to abnormal splicing of *IBKAP*

Spliced mRNA missing exon 20

Figure 3–1. Mutation in an intron leads to aberrant splicing, exon skipping, and familial dysautonomia. Correction uses kinetin, a plant cytokinin that suppresses this mutation. Note that abnormal splicing can be suppressed by kinetin (modified from Hims et al. 2002a, b).

sense sequences that target ESE sequences may be used to induce skipping of exons that contain mutations (Cartegni and Krainer 2003).

SIGNIFICANCE OF NUCLEOTIDE CHANGES IN CODONS

It is important to consider information on the significance of nucleotide changes that have been generated over the past several decades through the analysis of mutant proteins in diseases due to single gene defects.

In 1974, Grantham published a formula for assessment of the significance of substitution of amino acids. In assessing the difference between amino acids, he noted that the following key factors should be taken into account: the composition of side chains, the polarity of the amino acid, the molecular volume of the amino acid, and the ratio of noncarbon elements in end groups or rings to carbons in the side chains. The Grantham formula defines the degree of difference between two amino acids.

In specific Mendelian disorders, disease severity differs depending on the causative mutation. Severe disease is associated with deletions, insertions, splice site mutations, or substitutions that lead to termination of transcription (stop codons). In the case of mutations that lead to amino acid substitution, the Grantham scale can be used to determine the impact of a substitution. Other matrices used to assess similarity of different amino acids, homology of proteins, and the functional effects of specific amino acid substitutions include Blosum, developed by Henikoff and Henikoff (1992).

In 2003, Vitkup et al. published results of analysis of 4236 disease-causing mutations in 436 genes and analysis of 1037 synonymous and non-synonymous single nucleotide polymorphisms (benign SNPs). Results of their analyses provided further support for the earlier studies of Grantham. Both studies led to the conclusion that the probability that a nonsynonymous nucleotide mutation within an exon will lead to genetic disease is directly related to the interspecies evolutionary conservation at that site. Vitkup et al. concluded further that substitutions that decrease the solvent accessibility at a particular mutation site have a more pronounced effect. They determined that amino acid substitutions in the interior of a protein are more likely to have a functional effect than mutations in amino acids located on the surface of the protein. Interior mutations frequently alter protein stability. Vitkup et al. concluded that approximately 15% of disease-causing mutations occur at arginine residues. This is due to the occurrence of CG in the arginine codon. Mutations in the arginine codon can lead to substitution to a number of different amino acid residues, cysteine, glycine, histidine, lysine, leucine, methionine, proline, glutamine, serine, and tryptophan. These residues each have markedly different structural properties than arginine.

ARE SYNONYMOUS POLYMORPHISMS FUNCTIONALLY SIGNIFICANT?

There is evidence that synonymous polymorphisms may impact function. Nackley et al. (2006) published evidence for the functional significance of synonymous substitutions on the products encoded by the *COMT* gene. This gene encodes both soluble (S) and membrane-bound (MB) forms of catechol-*O*-methyltransferase through use of different transcription initiation sites. One downstream effect of the *COMT* gene products is modulation of pain sensitivity. Four SNPs occur in the *COMT* gene, one in the S *COMT* promoter region, and three in the MB *COMT* coding sequence: his62his C/T rs4633, leu136leu C/G rs4818, and val158met is a nonsynonymous polymorphism. The last polymorphism lowers enzyme activity. Nackley et al. noted that the promoter region polymorphism did not impact pain sensitivity. Individuals with the Val substitution at position 156 had different pain sensitivity depending on the occurrence of the other nucleotide substitutions within their haplotype. These investigators proposed that the synonymous nucleotide substitution altered the secondary structure of the mRNA and that this in turn impacts mRNA degradation and protein translation efficiency. They carried out computer analysis of secondary structure of the mRNA derived in each of the haplotypes. They also transfected individual clones, each representing a different haplotype, into PC12 cells and demonstrated a correlation between haplotype and enzyme expression. Nackley et al. performed site-directed mutagenesis and demonstrated that the stable stem loop structure of mRNA prior to splicing is affected by base pairing between the specific nucleotides 403C and 479G.

These data reveal that synonymous nucleotide changes lead to altered mRNA secondary structure and have a pronounced effect on enzyme activity.

Duan et al. (2003) investigated synonymous nucleotide variations in the *DRD2* gene. They noted that some of these variants had functional effects. They proposed that effects arose because specific synonymous variants led to decreased mRNA stability and translation. They demonstrated further that combinations of synonymous mutations may have functional consequences drastically different from those of isolated mutations.

NONSENSE MUTATIONS AND STOP CODONS:
THERAPEUTIC POSSIBILITIES

Nonsense mutations are single nucleotide mutations that change an amino acid code to a stop codon, so that the ribosomal machinery prematurely terminates synthesis of the polypeptide chain. Natural defenses against the effect of stop codon mutations and prematurely terminated polypeptide chains include pioneer translation and nonsense-mediated mRNA decay that involves degradation of the mRNA in which a premature termination codon is present. Pioneer translation differs from productive translation. In pioneer translation, the ribosome scans mRNA for the presence of premature termination codons.

Premature termination codons constitute the underlying defect in a number of genetic diseases such as Duchenne muscular dystrophy and cystic fibrosis.

There is now evidence that treatment with a small organic molecule, PTC124, can eliminate termination of translation induced by stop codon mutations (Schmitz and Famulok 2007). Gentamycin treatment can lead to skipping of stop codons. However, the concentrations of the drug required to achieve this lead to severe side effects. There is evidence from phase II clinical trials that PTC124 achieves stop codon skipping at concentrations that do not lead to side effects (Welch et al. 2007). Clinical trials have included patients with cystic fibrosis and Duchenne muscular dystrophy.

In a number of disorders, boosting protein translation by 1% to 5% will reduce or eliminate the disease manifestations. This is true in cystic fibrosis. There is therefore a drive to identify drugs that suppress premature translation termination that arises in consequence of nonsense mutations that convert an amino acid codon to a stop codon, UAA, UAG, or UGA. There is evidence that the nonsense mRNA may produce a functional protein if its decay rate can be altered.

A number of specialized factors are recruited to the site of premature stop codons to ensure nonsense-mediated mRNA decay. If these factors are in-activated, read through can take place. Welch et al. (2007) carried out high-throughput screens of 800,000 low molecular weight compounds that suppress UGA nonsense codons. These screens led to identification of the compound PTC124. This compound has no structural similarity to gentamycin. PTC124 promotes read through of three nonsense codons; however, it leads to more read through of UGA codons (Figure 3–2).

Welch et al. carried out studies on cultures of muscle cells from Du-chenne muscular dystrophy patients and studies on mdx mice (Duchenne muscular dystrophy models). They demonstrated that PTC124 increased dystrophin production, and levels of dystrophin-associated protein sarcogly-can were also increased. They demonstrated further that normal termination codons were not suppressed. This difference is likely based on mechanistic differences between normal and premature termination codons.

TRANSLATIONAL CONTROL OF GENE EXPRESSION

Variation in Translation of Beta-Secretase (*BACE1*)

The AUG methionine codon at the 5′ end of the mRNA transcript is most commonly the translation initiation codon. In some genes, there are a series of 5′ AUG codons. Six AUG codons occur in the 5′ leader sequence of the gene that encodes the beta-secretase *BACE1*, the beta-site beta-amyloid precursor

Translation of spliced mRNA, effect of premature stop codons, and treatment of PTC

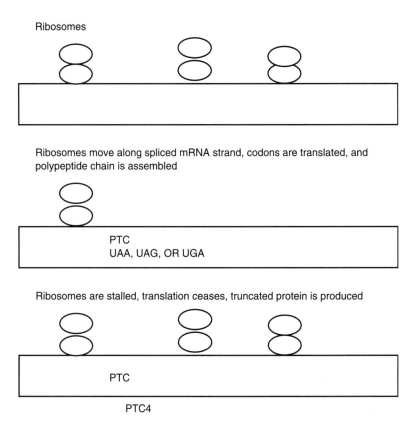

Ribosomes

Ribosomes move along spliced mRNA strand, codons are translated, and polypeptide chain is assembled

PTC
UAA, UAG, OR UGA

Ribosomes are stalled, translation ceases, truncated protein is produced

PTC

PTC4

Chemical compound PTC124 suppresses identification of PTC, ribosomes continue translation, polypeptide chain is produced

Figure 3–2. Mutations that generate premature stop codons lead to ribosome stalling and production of truncated polypeptides. The chemical compound PTC124 suppresses identification of premature stop codons and allows production of longer polypeptides (based on Welch et al. 2007).

protein cleaving enzyme 1. Zhou and Song (2006) demonstrated that ribosomes skip over some AUG sites and translation is usually initiated at the fourth AUG. They note that alterations in scanning and initiation of translation may alter *BACE1* gene expression and play a role in the pathogenesis of Alzheimer disease.

Translational Control of Expression of Glial Glutamate Transporter EAAT2

Glutamate is the major excitatory neurotransmitter. Following neurotransmission, glutamate is cleared from the synaptic cleft through the activity of glutamate transporters. The EAAT2 transporter is responsible for clearance of 90% of glutamate. Malfunction of glutamate clearance and accumulation of excessive glutamate in the synaptic cleft occurs in neurodegenerative diseases such as amyotrophic lateral sclerosis and Alzheimer disease. This malfunction occurs despite the fact that the levels of EAAT2 mRNA are normal and is apparently due to abnormal translation. Tian et al. (2007) reported that in normal cells a number of EAAT2 transcripts have long 5' untranslated regions and that these are less efficiently translated. The length of the 5' untranslated region may influence ribosomal function. They noted that many different cellular factors, including corticosterone and retinol, influenced which transcripts were most abundant in neurons.

ANALYSIS OF DOWNSTREAM EFFECTS OF GENE MUTATIONS: POLYGLUTAMINE EXPANSION AND INTERACTION WITH TRANSCRIPTION FACTORS

Huntingtin Polyglutamine Expansion, Transcription Factor Interactions

The mutant protein in Huntington disease (HD), huntingtin, contains an expanded stretch of polyglutamine residues. There is now evidence that the deleterious effect of this protein may be due to the fact that the polyglutamine residues interact with glutamine residues in transcription factors (Sugars and Rubinsztein 2003). Mutant huntingtin binds to transcription factors and transcription coactivators, leading to depletion in the activity of these factors.

Modulation of Expression of PGC1-alpha

Cui et al. (2006) reported that mutant huntingtin modulates the expression of the PGC-1alpha gene at the level of transcription. They concluded that mutant huntingtin associates with the chromatin surrounding the PGC1-alpha promoter and that this interaction limits the binding of the CREB and TAF4 transcription factors, leading to deficiency of PGC1α transcripts. PGC1-alpha is also sometimes given the symbol PPARGC1A (peroxisome proliferator-activated receptor gamma coactivator alpha).

PGC1-alpha is a transcriptional coactivator that regulates many genes, including those involved in mitochondrial biogenesis, oxidative phosphory-lation, and genes that encode electron transport proteins. Cui et al. noted that this may in part explain the finding of mitochondrial dysfunction in HD. These investigators reported that *PGC1-alpha* mRNA levels were decreased in caudate nucleus in HD patients but not in controls (Figure 3–3).

PGC1-alpha and Reactive Oxygen Species

St. Pierre et al. (2006) determined that *PGC1-alpha* is required for induction of many enzymes involved in detoxification of reactive oxygen species (ROS). They further demonstrated the *PGC1-alpha* null mice are more sensitive to the neurogenic effects of oxidative stressors.

Rohas et al. (2007) demonstrated that *PGC1-alpha* induces the expres-sion of enzymes involved in detoxification of ROS. They demonstrated further that *PGC1-alpha* and *PGC1-beta* normally regulate the ROS defense system in a number of tissues including brain. In *PGC1-alpha* knockout mice, they demonstrated increased sensitivity to oxidative stress. This leads to damage of dopaminergic cells in the hippocampus and in the substantia nigra. They reported that in *PGC1* null fibroblasts expression of the uncoupling protein 2 UCP2 was reduced to half and the production of components of the ROS defense system, including *SOD2*, catalase, and *GPX1*, were significantly re-duced. Importantly, the studies of Rohas et al. (2007) revealed that *PGC1-alpha* induces expression of anti-ROS genes, not only in the mitochondria but also in the cytoplasm (*SOD1*) and in the peroxisomes (catalase and glutathione peroxidases).

St. Pierre et al. (2006) concluded that *PGC1-alpha* induction may limit the defective mitochondrial function seen in a number of different neurode-generative disorders.

Weydt et al. (2006) analyzed the metabolic consequences of the HD mutation in transgenic mice. They noted altered response to stress, particularly to cold stress. They analyzed metabolism in brown adipose fat and determined that induction of uncoupling protein UCP1 was decreased in the HD mice. These defects were corrected by introduction of exogenous *PGC1-alpha*. Wheydt et al. further determined that reduced oxygen consumption occurs in the brain in HD mutant mice and that there is decreased expression of *PGC1-alpha* target genes in the striatum.

PGC1 coactivators dock with specific transcription factors. *PGC1-alpha* was first identified through its functional interaction with the nuclear receptor PPAR gamma. Finck and Kelly (2006) reported that *PGC1-alpha* regulates

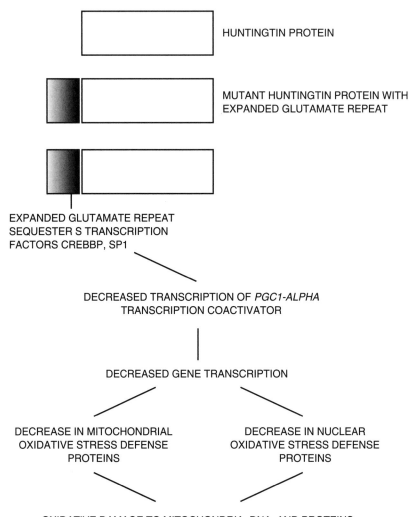

MUTANT HUNTINGTIN, DECREASED *PGC1-ALPHA* TRANSCRIPTION OXIDATIVE STRESS DAMAGE

HUNTINGTIN PROTEIN

MUTANT HUNTINGTIN PROTEIN WITH EXPANDED GLUTAMATE REPEAT

EXPANDED GLUTAMATE REPEAT SEQUESTER S TRANSCRIPTION FACTORS CREBBP, SP1

DECREASED TRANSCRIPTION OF *PGC1-ALPHA* TRANSCRIPTION COACTIVATOR

DECREASED GENE TRANSCRIPTION

DECREASE IN MITOCHONDRIAL OXIDATIVE STRESS DEFENSE PROTEINS

DECREASE IN NUCLEAR OXIDATIVE STRESS DEFENSE PROTEINS

OXIDATIVE DAMAGE TO MITOCHONDRIA, DNA, AND PROTEINS

Figure 3–3. The expanded glutamate repeat in Huntingtin protein sequesters transcription factors and decreases transcription of *PGC1-alpha*. This leads to decrease in transcription of oxidation defense proteins and to oxidative damage of DNA and proteins. One possible therapeutic intervention involves use of agents that enhance production of *PGC1-alpha* (based on Cui et al. 2006).

multiple pathways involved in cellular energy metabolism. *PGC1-alpha* interacts with the nuclear regulatory factors NRF1 and NRF2. These in turn regulate mitochondrial biogenesis and nuclear genes that play a role in mitochondrial function.

In a review of the findings on *PGC1-alpha* transcriptional regulation and abnormal mitochondrial function in HD, Ross and Thompson (2006) noted that mitochondrial toxins such as nitropropionic acid cause selective death of medium spiny neurons in the striatum of the basal ganglia. These major protection neurons are selectively affected in HD.

Possible Therapeutic Interventions: *PGC1-alpha* Induction

Hondares et al. (2006) examined *PGC1-alpha* transcription in response to thiazolidinediones, drugs that are used to treat diabetes mellitus type 2. They reported that these drugs induce expression of *PGC1-alpha* in white adipocytes. They attributed this induction to the presence of a response site to peroxisome proliferator activator receptor PPAR gamma. This response site is located in the distal promoter region of *PGC1-alpha*.

This raises the interesting question as to whether thiazolidinedione drugs may be useful in the treatment of neurodegenerative diseases, including HD (Bordet et al. 2006).

4

EPIGENETICS

Epigenetics may be defined as the study of changes in gene expression that are stable and potentially transmitted and do not entail a change in DNA sequence. The term *epigenetics* is sometimes used to refer to chromatin modifications and their consequent effects on the readout of genetic information. Epigenetic mechanisms include DNA methylation, covalent modification of histones (the histone code), and binding of proteins involved in chromatin remodeling. Chromatin remodeling plays a key role in gene expression since it determines accessibility of DNA to transcription factors (Jenuwein and Allis 2001).

Temporal and spatial changes in DNA methylation and chromatin state within an individual result in modification of gene expression. These changes may arise in response to metabolic or environmental factors.

CHROMATIN STRUCTURE

The basic unit of chromatin is the nucleosome. DNA strands are wound around the outside of beadlike structures called nucleosomes. A helix comprising approximately 165 base pairs of DNA is wound around a nucleosome bead. The bead is formed by histone subunits including H3 H4 tetramers and H2A H2B dimers. Histone H1 binds to the nucleosome and to linker strands of DNA that lie between the nucleosomes.

Chromatin remodeling plays a key role in gene transcription. Transcribed genes have an open chromatin structure; the intervals between nucleosomes are larger than in untranscribed genes. The open chromatin structure facilitates the binding of transcription factors to DNA, particularly to promoter regions of genes and to control elements. In untranscribed genes, nucleosomes are closer together and this tight packing prevents binding of transcription factors. Histone modifications impact chromatin secondary structure and influence binding of histones to other proteins (Jenuwein and Allis 2001). Histone proteins are composed of a globular region and a tail; amino acids in the tail undergo enzyme-induced modifications including acetylation, methylation, phosphorylation, and ubiquitination.

Chromatin remodeling is an energy-requiring process. A number of the protein complexes that bind to chromatin have ATPase activity and interact with ATP-rich compounds that provide energy for the remodeling process.

Histone acetyl transferases (HAT) transfer acetyl groups to the N terminal groups of histone tails. Acetylation of the N terminals of the histone tail leads to decreased interactions of histone with the phosphate backbone of DNA; this then leads to nucleosome relaxation. Nucleosomal relaxation facilitates access of transcriptional activators to DNA.

Histone deacetylases (HDAC) remove acetyl groups from histone N terminals; this leads to hypoacetylated chromatin and to silent genes. There are at least 11 different forms of HDAC. They occur in a complex with other proteins that act as transcriptional repressors. Histone deacetylases are inhibited by trichostatin and also by valproic acid and by a number of other small molecules that are discussed further below (p. 61).

DNA METHYLATION

In mammalian DNA, the major modified base is 5 methylcytosine (C); 2% to 5% of all cytosines are methylated. Methylated C occurs almost exclusively in cytosine guanine, CpG, dinucleotides. The promoter regions of genes are particularly rich in methylated CpG. Approximately 50% of genes have CpG islands in their promoter regions or first exons. Methylation of these islands inhibits gene expression. Methylation of DNA also occurs in noncoding repetitive DNA sequences and in retroviral elements that are incorporated into genomic DNA.

Enzymes that establish and maintain DNA methylation patterns include DNA methyltransferases, DNMT2, DNMT3a, and DNMT3b that catalyze methylation. DNMT3L acts by binding to other DNMT proteins and altering their DNA methylation activity. The methylation pattern can be transmitted during cell division. Maintenance methylase DNMT1 recognizes hemimethylated DNA at the replication forks in DNA and then methylate the new DNA strand. DNA methylation can directly inhibit the binding of transcription factors (Jaenisch and Bird 2003).

DNA methylation and histone modification are closely associated processes. Methylated DNA often binds protein complexes (methyl CpG binding proteins). Several proteins have affinity for methylated DNA through their methyl-binding domains. These include proteins that have methyl-binding domains MBD1, MBD2 and MBD3, and MECP2. Following binding to methylated DNA, through its methyl-binding domain, MECP2 recruits a complex that includes SIN3A (Jaenisch and Bird 2003).

These bound proteins recruit histone deacetylases. Enzymes responsible for histone deacetylation occur in the vicinity of repressed genes. In addition, the MBD proteins recruit ATP-dependent helicase, which alters chromatin structure and impacts the activity of RNA polymerase that is involved in DNA transcription. Deacetylation of histones via histone deacetylases leads to a closed structure that is not accessible to transcription factors and genes are silenced.

Acetylation of histone H3 and histone H4 leads to an open chromatin configuration; actively expressed genes are surrounded by acetylated histones.

IMPRINTING

Definition

Genomic imprinting is defined as the differential expression of two alleles at a specific gene locus based on their parent of origin. Bartolomei and Tilghman (1997) noted, "a handful of autosomal genes in the mammalian genome are inherited in a silent state from one of the two parents, and in a fully active form from the other, thereby rendering the organism functionally hemizygous for imprinted genes" (p. 493).

Genomic imprinting is unique to mammals. Imprinted genes are clustered within domains and imprint control regions regulate allele-specific expression.

Imprint control regions are commonly located in the 5′ gene regions though in some cases they reside in introns. Imprint control regions are usually marked by hypermethylation. Hutter et al. (2006) reported that imprinted genes more frequently contain random repeat arrays and intragenic CpG islands. They postulate that the latter may serve as promoters for antisense transcripts, in an allele-specific manner. There is evidence that RNA controls epigenetic phenomena. RNA may silence DNA by posttranscriptional mechanisms, such as binding of antisense RNA to sense RNA.

Imprinting and the Germline

After fertilization, genomes derived from both sperm and oocytes undergo erasure of lineage-specific methylation. Demethylation begins within hours of fertilization and continues during cleavage. During further developmental stages, sex-specific epigenetic patterns are established. Imprinting may be specific for a particular developmental stage. Imprinting may be tissue specific.

The very low success rate in cloning of mammals is likely due in part to aberrations in DNA methylation and imprinting associated with somatic cell nuclear transfer into egg cytoplasm (Rideout et al. 2001).

TECHNICAL DEVELOPMENTS FOR EPIGENETIC STUDIES

Technical developments include studies on the "DNA methylome." Analysis of cytosine methylation may be performed using bisulfite sequencing, as first described by Herman et al. (1996). However, this technique cannot be readily scaled up. Genomic studies are currently more frequently carried out by fractionating the genomic DNA samples into methylated and unmethylated portions using methylation-sensitive restriction endonucleases or antibodies, followed by sequencing and array analysis. Large-scale analyses of DNA and histone modification often involve chromatin immune precipitation using an antibody that binds to a chromatin-associated protein. The presence of a defined DNA segment within the immune precipitated chromatin is sought using sequencing or array analysis. Immune precipitation is also used to identify specific DNA-binding proteins and the DNA targets of those proteins (Goldberg et al. 2007). Such studies are, for example, used to analyze methylation patterns in different cell types and tumors. There are reports that distinct epigenetic signatures are associated with specific tumors (see p. 60).

IMPRINTING DISORDERS

Autosomal Disorders

Examples of autosomal disorders where imprinting plays a role are Prader-Willi syndrome, Angelman syndrome, and Beckwith-Wiedemann syndrome.

One hallmark of a disorder where imprinting plays a role is that the parent of origin of the molecular defect influences the phenotype. Furthermore, uniparental disomy is often associated with the disorder. In uniparental disomy, part or all of both chromosomes in an individual pair are inherited from the same parent. For example, in paternal uniparental disomy, part or all of both chromosomes in a specific pair are inherited from the father. In uniparental isodisomy, the same parental chromosome is present in two copies, due to failure of chromatids to segregate at the second meiotic division. In uniparental heterodisomy, two different chromosome homologs were derived from a single parent. This occurs due to failure of reduction division at the first meiotic division.

ABERRANT HYPERMETHYLATION OF TRINUCLEOTIDE REPEAT EXPANSION ON X CHROMOSOME

Repeat expansion and associated hypermethylation lead to reduced or absent expression of the FMR protein and to fragile X mental retardation. Methylation of the expanded CGG repeat is associated with hypoacetylation of histones and with condensation of chromatin. Coffee et al. (2002) demonstrated that the degree of acetylation of histone H4 in the FMR1 region chromatin is proportional to the size of the CGG repeat. They also reported increased methylation of histone H3 at lysine 4 and lysine 9.

DNA METHYLATION AND EPIGENETIC FACTORS IN CANCER

Changes in the methylation state of specific genes frequently occur in cancer cells. In some instances, hypermethylation of promoters of tumor suppressor genes leads to loss of expression of these genes. Methylation changes are often demonstrable in precancerous lesions, indicating that these changes occur early in oncogenesis (Jones 2002).

In cancer cells, oncogenes may be activated by changes in methylation. Activation may be due to hypomethylation of promoters or transfer of genes from hypermethylated chromosome regions (heterochromatin) to hypomethylated regions (euchromatin), through chromosome rearrangements such as translocations.

Drugs That Impact Methylation

New drugs for cancer are being developed that act by altering methylation states. For example, if the major change in a specific tumor is loss of tumor suppressor activity through promoter methylation, it may be possible to reactivate that gene through use of agents that alter methylation (Jones 2002).

Pharmacological treatments designed to reactivate tumor suppressor gene expression include use of DNA methyltransferase inhibitors such as azanucleotides that lead to DNA demethylation. An azanucleotide approved for use in the United States by the Food and Drug Administration for treatment of myelodysplastic syndrome is 5-aza-2-deoxycytidine (Decitabine). Treatment is apparently well tolerated by patients. Muller et al. (2006) reported that karyotypes were normalized in 30% of patients following treatment.

Histone Deacetylase Inhibitors in Cancer Treatment

In a variety of different tumor cell lines, including those established from erythroleukemia, bladder transitional cell carcinoma, and breast adenocarcinomas, the histone deacetylase inhibitor suberoylanilide hydroxamic acid (SAHA) induces apoptosis and cell differentiation (O'Connor et al. 2006). The mode of action of HDAC inhibitors in a number of cancers is apparently related to their impact on gene transcription.

O'Connor et al. reported that in lymphoma cells, proto-oncogenes such as BCL6 are abnormally activated and suppress transcription of genes involved in cytodifferentiation. Reversal of this activity through acetylation (i.e., through inhibition of deacetylation) promotes normal differentiation and suppresses malignancy.

Phase 1 clinical trials revealed that this drug is well tolerated. O'Connor et al. (2006) reported that Phase 2 trials are ongoing for use of SAHA in hematological tumors and in solid tumors.

HDAC INHIBITORS IN THE TREATMENT OF NEURODEGENERATIVE DISEASES

Studies in models of Huntington disease have revealed that transcriptional dysregulation frequently occurs. Drummond et al. (2005) reported that histones are not the only proteins modified by acetylation. Many proteins undergo posttranslational modification involving acetylation and deacetylation. These include transcription factors and mediators of signal transduction.

Use of HDAC inhibitors leads to inhibition of activity of HDAC6 and increased acetylation of tubulin (Dompierre et al. 2007). This leads to increased microtubule-based transport of the neurotrophic factor BDNF. Dompierre et al. demonstrated further that in Huntington disease tubulin activation is reduced. This reduction can be reversed by treatment with HDAC inhibitors, leading to increased intracellular BDNF.

USE OF HDAC INHIBITORS TO TREAT THALASSEMIA

In a mouse model of beta-thalassemia, Cao et al. (2005) demonstrated that hydroxamic acid derivatives of short-chain fatty acids, specifically propionyl and butyryl hydroxamates, increased fetal globin production and increased erythropoiesis. They concluded that these compounds may be useful in the treatment of beta-chain hemoglobinopathies.

5

COMPLEX DISEASES: MOLECULAR GENETIC CONTRIBUTIONS TO ELUCIDATION OF ETIOPATHOGENESIS OF LATE-ONSET NEURODEGENERATIVE DISEASES

In complex diseases, genetic factors and environmental factors both play a role, and disease manifestations often occur later in life. Single gene defects may play the most important role in some forms of these diseases; often these constitute the rare forms of that disease and age of onset is somewhat earlier. Identification of genes involved in these rare earlier onset forms has provided great insight into the disease processes, and these insights are also relevant to the pathogenesis of the later onset forms. This is true, for example, in Parkinson disease (PD) and Alzheimer disease (AD). More often there is evidence that genetic susceptibility, determined by a number of different genes, interacts with specific environmental factors to lead to late-onset disorders. The specific changes or alteration in the genes that influence risk of late-onset disorders may occur with relatively high frequency within the population and be classified as polymorphisms.

In working toward prevention of development of disease symptoms, it is obviously most beneficial to consider environmental factors and whether or not these may be modulated. In development of therapies, it is critical that basic disease mechanisms and steps in the disease pathogenesis be understood. Understanding the pathogenesis of a disease often requires that much research be done to investigate the nature and function of "normal" molecules and that the pathways in which these molecules participate be examined.

In addition to genomics and availability of gene sequence, advances in proteomics play an important role. Proteomic studies lead to an understanding not only of the primary structure of a protein but also of its folding and three-dimensional structures. Analysis of protein modification and its longevity in the cell and the mode of degradation are also important. The impact of the milieu in which the protein resides must also be taken into account. Increasingly, oxidative stress within the cell is implicated in aberrant protein modification and disturbances in cellular protein processing in chronic neurodegenerative diseases. There is also a growing body of evidence that oxidative stress modifies DNA and that this has deleterious effects on gene function.

ENVIRONMENTAL FACTORS IN LATE-ONSET DISORDERS

Superoxide and Oxidative Stress

Superoxide is a free radical of oxygen that has an unpaired electron, O_2^-. It is unstable and converts to peroxide O_2^{2-} with two free electrons and to H_2O_2 (hydrogen peroxide). Superoxide is toxic. It is produced in the body, for example by leukocytes, and serves to destroy pathogens. Superoxide is produced through the activity of NADPH (nicotinamide adenine dinucleotide phosphate [reduced]) oxidase. It may also be produced as a by-product of mitochondrial respiration. Mitochondria apparently do not produce considerable quantities of superoxide under normal conditions. However, they produce superoxide in the presence of respiratory complex 1 inhibitors (St.-Pierre et al. 2002).

Oxidative stress is defined as the condition that results when there is an accumulation of harmful oxygen radicals and a decrease in antioxidant substances. Vitamins C, A, and E and lycopenes are natural antioxidants. Glutathione acts as a cofactor for peroxidase enzymes that metabolize H_2O_2. Glutathione may also participate in nitric acid homeostasis. Glutathione is a tripeptide formed by linkage of glutamine to cysteine and glycine.

Accumulated oxidative stress resulting from a gradual shift in the redox status of tissues is now considered to be a key mechanism underlying the aging process. Calorie-restricted feeding, an experimental protocol to extend survival and delay aging in rodents, is recognized to slow the rate of accrual of age-related oxidative stress (Merry 2004).

Superoxide binds to and inactivates iron sulfur cluster enzymes. Through this mechanism it exerts its toxic effect, since many of these enzymes are involved in metabolism. Iron release facilitates generation of the highly re-active hydroxyl radical HO_2, in a reaction known as the Fenton reaction: Fe^{++} reduces H_2O_2 to HO_2.

Oxidative damage to DNA may occur through modification of purine and pyrimidine base and through modification of the sugar residues. Damage to the sugar residues of DNA may lead to strand breakage and subsequent release of bases (Gracy et al. 1999). Oxidative modification of proteins leads to protein cross-linking, peptide fragmentation, and amino acid conversion. These amino acid changes may in turn lead to protein conformational changes. Superoxide and nitric oxide interact to give rise to peroxynitrite. This may react with methionine or tyrosine and secondarily affect proteins. Oxidative damage to proteins may also impact the process of ubiquitination and proteosomal degradation (Jenner 2003).

Superoxide radicals may interact with carbohydrates and with lipids, giving rise to ketoaldehydes, hydroxynonal, and malondialdehyde.

Reactive oxygen species (ROS) damage neuronal cell membranes, leading to calcium influx. This in turn may activate calmodulin-induced activity of nitric oxide synthase. Nitric oxide and superoxide interact to form peroxynitrite. Nitration of amino acids, particularly tyrosine, then impacts function of proteins involved in signal transduction.

There is evidence that oxidative stress plays a key role in the pathogenesis of PD. There is evidence that amyloid basic protein that accumulates in AD generates ROS (Gibson et al. 2004).

Oxidative DNA Damage and Somatic Mutation: Their Role in Huntington's Chorea

Huntington disease (HD) is caused by expansion of the polyglutamine-encoding repeat in the gene that encodes the protein huntingtin. Repeat expansions to greater than 36 repeat units are associated with disease symptoms during adult life. The most striking pathological changes in HD occur in the striatum, particularly in the medium spiny neurons. A number of investigators have reported somatic variation in the number of CAG repeats in HD patients. Shelbourne et al. (2003) reported that the somatic instability of the CAG repeat impacted tissue-specific pathology in HD. Their studies on a mouse model of HD demonstrated dramatic age-dependent expansion of the CAG repeat in the striatum. In studies on human brain, there is evidence that somatic expansion of the HD repeat occurs first in the striatum. In one patient, they noted that the median CAG expansion number was 41 while in the same patient the expansion number in the striatum was 1000 CAG repeats.

The R6/1 strain mice are transgenic for human exon 1 of the HD gene that carries an expanded CAG repeat. Studies in these mice revealed that the CAG repeat is initially stable in size but begins to expand in midlife (at about 11 weeks). The repeat length then continues to expand with age (Cummings and Zoghbi 2000).

Kovtun et al. (2007) investigated whether oxidative damage played a role in HD repeat expansion in the R6/1 strain. They demonstrated that oxidative DNA changes occurred in the liver and brain of these mice and that these changes increased with age. The specific oxidative DNA compounds they detected included 8-oxo-G (7,8-dihydro-8-oxyguanine), 5-hydroxyuracil, 5-hydroxycytosine, and formamidopyrimidine. Since the presence of oxidative DNA damage may indicate decreased capacity to repair DNA damage, Kovtun et al. measured DNA repair capacity in tissue. They determined that this was intact. Oxidized bases are repaired by a DNA glycosylase that cleaves the C1 glycosidic bond. This is followed by removal of ribose phosphate and generation of a single-strand DNA break, SSB. They noted that oxidative DNA damage coincided in time with CAG repeat expansion.

In HD fibroblasts from patients, Kovtun et al. induced oxidative changes with H_2O_2, one of the main compounds generated in mitochondria from the superoxide anion radical, O_2^-. Oxidative changes thus induced led to CAG repeat expansion and to single stranded DNA breaks. They concluded that CAG repeat expansion occurred in the process of repair of single stranded breaks.

Kovtun and coworkers (2007) then carried out studies on the enzymes that initiate break excision repair. A key enzyme in this process is the glycosylase OGG1 that recognizes and removes 8-oxo-G. They found that in mice resulting from the cross of OGG1-deficient mice with the R6/1 HD mice, expansion of the CAG repeat was significantly delayed. Removal of a damaged base requires OGG, and an endonuclease APE1 (apurinic apyramidinic endonuclease) and polymerase B. On a random DNA template, polymerase B added a single nucleotide and repaired the break. However, on CAG templates, OGG1/APE1 cleavage and polymerase B gap filling and synthesis generated longer products, and addition of three nucleotides was favored. There was evidence that OGG1-mediated base excision repair initiated expansion through strand displacement, slippage, and gap filling. Kovtun et al. postulated that the specific role of OGG1 in this process may indicate that within the CAG site 8-oxo-guanine DNA damage is most common. They noted further that somatic expansion of the CAG repeat did not require cell division.

These studies reveal the importance of oxidative DNA damage and single stranded DNA break and repair in the generation of HD. Kovtun and coworkers emphasized that this process is likely also important in other forms of neurodegeneration.

Mitochondria and Their Role in Late-Onset Diseases

Dietary calories are utilized by mitochondria to produce energy through oxidative phosphorylation. In this process, ROS are generated. Wallace (2005) postulated that mitochondrial dysfunction plays a key role in late-onset disorders such as AD, PD, and type 2 diabetes mellitus. Dietary caloric intake impacts mitochondrial function, and this impact may be more profound in individuals who carry specific inherited mitochondrial polymorphisms. Wallace noted that the rapidly increasing incidence of late-onset disorders must be due to environmental factors. He postulates that the most significant change in the environment is dietary, with increased caloric intake being the major factor. Regional population differences exist in mitochondrial DNA sequences, and these differences have permitted adaptation to different environmental conditions in the past. Frequently, adaptations involved biological accommodation to weather conditions and dietary shortages, which no longer pertain to the populations who carry the mutations.

Wallace (2005) presented evidence that mitochondrial damage dysfunction and the rate of mitochondrial DNA mutation is affected by ROS. Furthermore, he noted that ROS production is a function of caloric availability.

The mitochondrial DNA encodes genes for 13 polypeptides required for energy generation through oxidative phosphorylation. In addition, mitochondrial DNA encodes genes for two species of ribosomal RNA and 22 transfer RNAs. Approximately 1500 genes within nuclear DNA encode proteins that are imported into mitochondria.

Wallace (2005) emphasized the connection between mitochondrial oxidative phosphorylation, production of ROS, and initiation of cellular apoptosis. Apoptosis involves the mitochondrial permeability transition pore mtPTP. Opening of this pore permits release of several pro-apoptotic proteins including mitochondrial cytochrome C. Activation of cytosolic caspase 9 through function of mitochondrial cytochrome C leads to degradation of cellular protein.

In tightly coupled mitochondria, electron transport through complexes I to V is efficient and leads to high rates of ATP synthesis. In the absence of exercise or in the presence of excess calories, electron transfer is stalled and electrons may be donated to molecular oxygen O_2, resulting in the generation of the superoxide radical O_2^-. This is then processed by superoxide dismutase enzymes to produce H_2O_2, hydrogen peroxide. This in turn may be metabolized by glutathione peroxidase or by catalase. It may, however, give rise to highly reactive hydroxyl radicals. Reactive oxygen species particularly affect the iron sulfur centers of mitochondrial enzymes, thus impacting enzyme function.

Wallace (2005) noted that mitochondrial DNA has a high mutation rate. This may be attributed to the effect of ROS. Mitochondria are present in thousands of copies per cell. Mutant and normal mitochondrial DNA species replicate and may subsequently segregate to different cells in different ratios. Mitochondrial mutations include rearrangements, particularly insertions or deletions, and missense mutations.

Wallace proposed that mitochondrial mutations contribute to the aging process. He postulated that there are four steps in the pathway of mitochondrial contribution to aging. These include damage to mitochondria by ROS; mitochondrial DNA point mutations and rearrangements; amplification of mutant mitochondrial DNA; and activation of the mitochondrial permeability transition pore mtPTP, through which pro-apoptotic proteins are released. Wallace reported that the level of mitochondrial DNA rearrangements is greatly increased in the age-related degenerative disease AD.

Lopez-Lluch et al. (2006) determined that calorie restriction induces a factor that plays a key role in mitochondrial proliferation, through its effect on mitochondrial biogenesis via the peroxisome proliferation activated receptor coactivator 1 alpha (PPARGC1A), also known as PGC1alpha.

Mitochondria and Diabetes Mellitus

Wallace (2005) emphasized the mitochondrial etiology for type 2 diabetes. He noted that individuals with a partial OXPHOS (oxidative phosphorylation) deficit often have a limited capacity for carbohydrate utilization. With increased caloric intake, there is chronic mitochondrial oxidative stress and increased ROS production. High caloric intake also results in high glucose levels that stimulate insulin production. Mitochondrial function is often downregulated in diabetes mellitus. This is often associated with altered levels of peroxisome proliferation activated receptor gamma (PPARG), a regulator of mitochondrial biogenesis. Interestingly, in two different populations, the Danish and the Pima Indians, type 2 diabetes is associated with specific polymorphisms of PPARG.

Wallace noted that one form of early-onset autosomal dominant forms of diabetes (MODY, maturity-onset diabetes of the young) is associated with polymorphism in the transcription factor HNF1 (hepatic nuclear factor 1), which regulates transcription of nuclear encoded mitochondrial genes.

EVIDENCE FOR OXIDATIVE STRESS IN ALZHEIMER DISEASE, PARKINSON DISEASE, AND AMYOTROPHIC LATERAL SCLEROSIS

Oxidative Stress, Oxidative DNA Damage

Evidence of oxidative stress in AD includes increases in lipid peroxidation and in DNA and protein oxidation products, 8-oxo-G and protein carbonyls. The brain is abundant in oxygen. It consumes 20% of the body's oxygen though it accounts for 2% of body weight (Smith et al. 2007). Furthermore, with aging the brain concentration of metals such as iron and copper increases. These metals contribute to generation of ROS. The oxygen radical–mediated reactions result in lipid peroxidation derivatives such as electrophilic aldehydes; these radicals have a longer half-life than free radicals. DNA oxidation adducts occur in the nucleus and in mitochondria.

In AD, elevated concentrations of protein carbonyls are found in the hippocampus and medial temporal gyrus and in the frontal lobe; nitrotyrosine and dihydrotyrosine cross-linked protein concentrations are elevated. Reduced levels of antioxidants, such as thioredoxin and glutathione transferase, occur in the AD. The electrophilic aldehydes are inactivated through conjugation with glutathione. Elevated levels of conjugated glutathione and of the enzyme superoxide dismutase occur in the AD brain and are indicative of increased free radical attack. Superoxide dismutase catalyzes the conversion of superoxide O_2^- to H_2O_2. This in turn is converted to H_2O through the activity of the enzyme catalase. Smith et al. (2007) reported that amyloid beta peptide (Abeta) has strong reduction potential. It reduces Fe^3 and Cu^2 to yield free radicals. They reported evidence that metal ions play a key role in oxidative stress and amyloid beta toxicity. These investigators noted that antioxidants most likely are of therapeutic value as prophylactic substances. In addition to vitamin C and vitamin E, bioflavanoids, melatonin, and curcumin are of possible value. Orally administered iron chelators are in clinical trials. These include clioquinol, a quinidine derivative first used as an antibiotic.

The enzyme NADPH oxidase is activated by amyloid beta and its expression is elevated in AD brains. It plays a role in generation of superoxides. It is best known as the enzyme present in immunocompetent cells that is responsible for generation of the superoxide burst required for bacterial killing (Canevari and Clark 2007).

Repair of Oxidative DNA Damage

Enzymes that counteract oxidative DNA damage include glycosylase and MTH1 (mutH homologue). Oxidized purine nucleoside triphosphates such as 8-oxo-2 deoxyguanosine triphosphate and 2-hydroxy-2 deoxyadenosine triphosphate are hydrolyzed to monophosphate forms by MTH1. This prevents oxidized radicals from being present in DNA during replication and serves as an important repair mechanism for nuclear and mitochondrial DNA. OGG1, 8-oxoguanine glycosylase and 2-hydroxyadenine glycosylase are involved in cleavage of oxidized nucleotides from the DNA chain.

Levels of 8-oxoguanine are increased in mitochondria in PD, and Nakabeppu et al. (2007) postulated that this increase may play a role in loss of dopamine neurons. We discuss below (pp. 74–77) evidence from environmental and genetic studies that oxidative stress and mitochondrial dysfunctions play a key role in PD.

Genetic variation in 8-oxoguanine glycosylase may play a role in the neurodegenerative disease amyotrophic lateral sclerosis (ALS), which affects motor neurons. Coppede et al. (2007) reported a significant association between the cysteine allele at position 326 in the protein in males with ALS.

MECHANISMS TO REMOVE DAMAGED OR AGGREGATED PROTEINS: AUTOPHAGY AND THE UBIQUITIN PROTEOSOME PATHWAY

In eukaryotes, mechanisms exist to remove damaged proteins and to avoid pathological protein accumulation. A number of late-onset diseases are characterized by the excess accumulation of specific proteins. Intracellular pathways involved in the degradation of proteins include lysosomal degradation (autophagy) and the ubiquitin proteosome system.

The Ubiquitin Proteosome Pathway

Hershko and Ciechanover (1998) proved that ubiquitination is an energy-requiring process ATP-dependent system in which a subset of intracellular proteins are targeted with the small protein ubiquitin and thus marked for degradation. In the initial steps in the pathway, individual ubiquitin molecules are conjugated and attached to a substrate through the activity of three enzymes that act in concert. These include E1 ubiquitin activating enzyme, E2 ubiquitin conjugating enzyme, and E3 ubiquitin protein ligase. Ubiquitin ligase E3 catalyzes the transfer of the ubiquitinated proteins to the proteosome for degradation. There are few ubiquitin-activating enzymes. There are approximately 50 different enzymes with E2 ubiquitin conjugating activity and approximately 1000 different enzymes with E3 ubiquitin ligase activity. Ubiquitin release and recycling require the activity of a specific enzyme, ubiquitin carboxyl-terminal esterase L1.

The inclusions that occur in neurodegenerative diseases are often poly-ubiquitinated and serve as evidence that the ubiquitin proteosome system has failed. There is evidence that this system becomes less efficient with aging.

Protein Folding and the Associated Molecular Chaperone System

Protein folding and the molecular chaperone system that enhances protein folding become less efficient with aging (Huang and Ardsley 2006). Abnormalities in the proteins may predispose to misfolding and precipitation, and, for example, the A53T mutation in alpha-synuclein predisposes to misfolding and aggregation in one form of parkinsonism. The polyglutamine expansion in HD predisposes to misfolding; amino acid mutations in Tau lead to misfolding and formation of fibrillar aggregates in some forms of dementia. There is an ongoing debate as to whether the protein aggregates lead to cell death or whether neuronal damage precedes deposition of the proteins. Culture systems for cells containing specific inclusion-promoting mutant proteins, such as huntingtin and synuclein, represent tools for analysis of compounds that enhance or decrease aggregation.

Chaperone Mediated Autophagy

Autophagy is the lysosome-mediated pathway of protein degradation. In this process, membranes form around the intracellular target molecules and move them toward the lysosome. There is now evidence that the ubiquitin proteosome pathway and the autophagy pathway interact. Pandey et al. (2007) demonstrated that in *Drosophila* mutants where the proteosome system was defective, autophagy acted as a compensatory mechanism. Furthermore, they demonstrated that histone deacetylase 6 (HDAC6), which interacts with ubiquitinated proteins, was essential in autophagy. Their studies demonstrated that overexpression of HDAC6 suppressed degenerative phenotypes in other mutants associated with protein aggregation. These included amyloid beta, which accumulates in AD. There is evidence that autophagy and oxidative stress may be linked.

LATE-ONSET NEURODEGENERATIVE DISEASE: PARKINSON DISEASE

Parkinson disease serves as an important model of a disorder that is complex in its etiology, involving genetic and environmental factors. The genetic contributors can be major or minor. Both autosomal dominant and autosomal recessive forms of the disorder are present, and mitochondrial DNA mutations, inherited and sporadic, play a role.

Parkinson disease is also important in illustrating how basic science research is required to understand the pathogenesis of the disease and to devise treatments. Treatments may be devised to impact the symptoms of the disorder. The hope is, however, that it may be possible to specifically impact disease processes by first understanding the pathogenesis.

Symptoms

Idiopathic PD is characterized by a specific motor phenotype including tremor, rigidity, slowing of movement or absence of movement, and postural instability. It is also associated with a distinctive neuropathology, including substantial loss of dopaminergic neurons from the substantia nigra and the presence of alpha-synuclein positive inclusions in cell bodies and processes of specific neurons, particularly in the brain stem, Lewy bodies, and Lewy neurites.

Litvan, Chesselet, et al. (2007) and Litvan, Halliday, et al. (2007) noted that genetically determined Parkinson syndromes may resemble idiopathic PD but they often have distinctive features. In some of these disorders, alpha-synuclein deposits do not occur. In others, alpha-synuclein aggregates occur in the absence of destruction of dopaminergic neurons, for example, in dementias with Lewy bodies. They note that mild parkinsonian symptoms may occur in the elderly as a result of loss of dopaminergic neurons.

The specific location of Lewy bodies within the brain may vary. The occurrence of these lesions in the brain stem leads to some of the characteristic symptoms of PD. Parkinson disease with dementia is associated with cortical Lewy bodies. Lewy bodies may also occur in autonomic and peripheral neurons. Litvan, Chesselet, et al. (2007) and Litvan, Halliday, et al. (2007)

stated that the effect of Lewy bodies on neuronal function is not clear. However, there is co-occurrence of Lewy body–filled cells and neuronal cell loss.

Etiologic Factors

Specific etiologic factors may include single gene defects predisposing to forms of PD with Mendelian inheritance, or exposure to a specific toxin or a specific infectious agent. Age is the most strongly associated factor. Environmental factors that play a role include exposure to pesticides, herbicides, and fungicides.

Genetic Studies

Linkage analysis and positional cloning have led to identification of genetic defects leading to PD. Forms of PD due to single gene defects usually have an earlier age of onset than idiopathic PD. Autosomal dominant inherited forms of the disease are caused by mutations in alpha-synuclein (Park1), and in *LRRK2* leucine rich repeat kinase (Park8). In a few families with autosomal dominant PD, the alpha-synuclein gene SNCA was found to be duplicated or triplicated (Fuchs et al. 2007). Autosomal recessive inheritance of PD occurs in association with mutations in parkin that has ubiquitin carboxyterminal esterase activity, and plays a role in the release and recycling of ubiquitin from proteins. Autosomal recessive inheritance of PD also occurs with mutation in *PINK1*, PTEN-induced putative kinase 1 (Park6). Kitada et al. (2007) reported that *PINK1* plays a critical role in the release of dopamine and in neuromodulation and synaptic plasticity in the nigrostriatal circuit. Absence of *PINK1* impairs release of dopamine. They also suggest that altered dopamine physiology may play a role in the degeneration in this region. DJ1 mutations lead to autosomal recessive forms of PD (Park7). DJ1 plays an important role as an antioxidant and is discussed further on page 74.

The gene mutations that lead to early-onset PD that follows a Mendelian pattern of inheritance are rare. Klein and Schlossmacher (2006) noted that all known monogenic forms combined explain about only 20% of early-onset PD and less than 3% of late-onset PD at best. Furthermore, there is a high degree of phenotypic overlap between different monogenic forms of PD. There is considerable clinical heterogeneity among patients with mutations in the same PD-predisposing gene.

LRRK2 mutations account for 5% to 18.7% of familial PD and have also been found in some forms of sporadic PD. There is evidence that Ashkenazi Jewish patients and Arab patients with sporadic PD are more likely to have mutation in this gene than are patients from other ethnic groups.

Specific genetic polymorphisms in the synuclein-encoding gene are associated with an increased risk of PD (Mizuta et al. 2006). There is some evidence that heterozygotes for Gaucher disease who carry mutations in the glucocerebrosidase gene may be at increased risk for PD (Aharon-Peretz et al. 2004; Goker-Alpan et al. 2004).

Factors under consideration for research aimed at elucidating the etiology of sporadic PD are the concentrations of regulatory proteins, the specific functional capacity of molecular chaperones that influence alpha-synuclein folding, and the role of oxidative stress.

Oxidative Stress and Aberrant Mitochondrial Function in Parkinson Disease

There is growing evidence from environmental and genetic studies that oxidative stress is important in the etiology of PD. Giasson et al. (2000) reported that alpha-synuclein protein present in Lewy bodies showed evidence of oxidative and nitrative damage. The DJ1 protein that is mutated or deleted in a subset of families with autosomal recessive PD plays a key role as an antioxidant in neurons. Zhou et al. (2006) reported that DJ1 is very susceptible to oxidation at its cysteine 106 residue and that oxidation of DJ1 cysteine 106 may act as an oxidative stress signal. Wild-type DJ1 exists as a stable dimer. Overoxidation of DJ1 and specific mutations of DJ1 lead to destabilization of the dimer. The wild-type DJ1 dimer and oxidized DJ1, 2ODJ1, act as molecular chaperones that facilitate protein folding (Hulleman et al. 2007). Oxidized DJ1, 2ODJ1, prevents fibril formation of alpha-synuclein. However, on further oxidation of additional cysteine residues and of methionine residues in DJ1, overoxidation DJ1 forms are no longer capable of preventing fibril formation (Figure 5–1).

Abnormal iron accumulation in the substantia nigra within dopamine neurons is a prominent feature of PD. There is evidence that iron induces oxidative stress in the dopaminergic neurons (Mandel et al. 2004). It is also important to note that a number of parkinsonism-inducing toxic substances impair mitochondrial function and alter redox balance (Litvan, Halliday, et al. 2007).

There is evidence that excess of iron in the diet in newborn mice and excess iron in media in which neuroblastoma cells are cultured alters parkin activity, that is, activity of the enzyme involved in ubiquitin release and recycling.

Key factors in the etiology of sporadic PD are increases in ROS leading to impaired mitochondrial function and particularly compromised activity of mitochondrial complex 1. Exposure to pesticides that inhibit mitochondrial

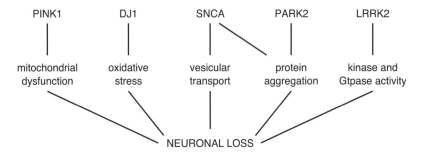

Figure 5–1. Identification of genes that play a role in forms of Parkinson disease that show Mendelian inheritance have shed light on pathogenetic mechanisms that also have relevance for sporadic forms of the disease.

complex 1 may lead to PD (van der Walt et al. 2003). MPTP, a contaminant in morphine synthesis, is concentrated in dopamine neurons through transport via the dopamine transporter. There it inhibits complex 1 activity, eventually leading to death of dopaminergic neurons. Rotenone causes complex 1 inhibition and PD. The herbicide paraquat and the fungicide maneb (a thiocarbamate derivative) cause dopamine neuron degeneration.

Interestingly, alpha-synuclein aggregation occurs not only in familial PD but also in sporadic PD, and its increased accumulation and aggregation occur in response to complex 1 deficiency (Sherer et al. 2003) and also in response to oxidative stress (Ischiropoulos and Beckman 2003). Synuclein can take on different conformations; it may be monomeric, fibrillar, amyloid-like, or Lewy body–like. Fibril formation is influenced by the degree of molecular nitration.

Mitochondrial DNA Deletions and Enzyme Function

Bender et al. (2006) carried out a study to determine the frequency of somatic mitochondrial DNA mutations in the substantia nigra (SN) in relation to aging and to parkinsonism. They also carried out histochemical analyses of cytochrome C oxidase and succinate dehydrogenase to determine the presence of mitochondrial respiratory chain defects. They determined that a significantly greater proportion of COX-deficient neurons occurred in individuals with PD as compared to normal controls. Analysis of mitochondrial DNA revealed that in SN neurons with COX deficiency there was a high frequency of deletions in mitochondrial DNA. However, studies on SN neurons with normal COX activity also showed high levels of deletion in the mitochondrial DNA. The latter showed a correlation with aging. However, there were statistically significant differences in the frequency of mitochondrial DNA (mtDNA) deletions between non-PD aged individuals and PD aged individuals. Bender et al. (2006) noted that deletions of mitochondrial DNA may be due to oxidative stress. SN neurons have a high oxidative capacity. They noted too that mtDNA deletions could be due to defects in mitochondrial DNA replication, specifically due to DNA polymerase (POLG) defects. Clonal expansion of cells with deleted mitochondria may explain their predominance. There are previous reports of POLG mutations associated with PD (Luoma et al. 2004).

There is evidence that calorie-restricted feeding lowers the rate of accumulation of superoxide radicals in mitochondria and slows oxidative damage in mitochondria (Merry 2004).

Greene (2006) carried out analysis of gene expression profiles of midbrain dopamine neurons. PD symptoms result from progressive degeneration of dopamine neurons, particularly in the ventral and lateral substantia nigra. Transcripts of genes related to energy metabolism were normally 2 to 50 times more abundant in the SN dopamine neurons than in other brain regions. The author states that the findings are consistent with previous observation on the susceptibility of SN dopamine neurons to inhibitors of complex 1 of the mitochondrial electron transport system. This led to the hypothesis that SN dopamine neurons are normally close to maximal metabolic capacity. Dopamine neurons that are less prone to damage (e.g., neurons in the ventral tegmental area of the midbrain) are higher in their expression of neuropeptides that may serve as neuroprotective factors.

Environmental Toxins and Destruction of Substantia Nigra

Toxins such as MPTP act to destroy SN cells by impairing the mitochondrial respiratory chain. MPTP (1-methyl 4-phenyl 1236 tetrahydropyridine) is a pethidine derivative that is sometimes a contaminant in illicit drugs. It is a potent inhibitor of mitochondrial complex 1. Paraquat is a structural analog of MPTP. It is used as a nonspecific weed killer (herbicide). Rotenone is a naturally occurring insecticide that is also used for killing fish, derived from the roots of several tropical and subtropical plant species belonging to the genus *Lonchocarpus* or *Derris*.

An atypical form of PD often associated with supranuclear palsy occurs in the French West Indies, and there is a postulated link with consumption of herbal tea and fruits from the Annonaceae family (*Annona muricata* and *Annona squamosa*, paw-paw) which contain neurotoxic benzyltetrahydroisoquinoline alkaloids. Among these alkaloids, acetogenins are known to induce dopaminergic cell loss in the substantia nigra (Caparros-Lefebvre and Lees 2005).

Therapy

Current therapy in PD is aimed at replacing dopamine. Other therapies being tested in model systems include strategies aimed at reducing alpha-synuclein synthesis and aggregation. Neuroprotective modifiers are also being evaluated.

ALZHEIMER DISEASE

Symptoms

Early features of AD include reduced ability to learn new information and to encode new memories. The loss of cognitive abilities is progressive and the impairment may involve language and visual-spatial skills and behavior. Clinical evaluation can only yield a probable diagnosis of AD. It is important to distinguish between AD disease and vascular dementia. Diagnosis of AD disease is made on the basis of neuropathology, and currently brain PET scans may give significant information on neuropathology. Onset of AD usually occurs after 65 years of age. Onset of symptoms described above before the age of 65 years is referred to as presenile dementia (Rocchi et al. 2003).

Neuropathology

Findings include the presence of amyloid plaques that occur outside neurons and neurofibrillary tangles that occur within neurons. Amyloid plaques are composed primarily of amyloid beta protein (Abeta) that is derived from amyloid precursor protein (APP). Abeta deposition occurs in plaques and blood vessel walls. There is often a marked inflammatory response in the vicinity of the Abeta deposit. Neuronal loss is often prominent and shrinkage of gyri also occurs. The protein Tau is also abnormally deposited in the brain in AD; it occurs primarily in the neurofibrillary tangles. Selkoe (2002) postulated that the subtlety of the early symptoms indicate that some discrete interruption of synapse function may be the underlying factor. There is growing evidence that subtle alterations of hippocampal function precede neuronal degeneration.

Amyloid Biology and Alzheimer Disease

Amyloid precursor protein

The extracellular deposits that accumulate in AD are composed primarily of 39–42 amino acid peptides derived from APP. In recent years, the molecular biology of APP has been intensively studied in AD (Rosenberg 2006). Early evidence of the importance of APP in the etiology of AD came from studies on individuals with trisomy 21 and Down syndrome. These individuals often develop signs of AD by the third decade of life.

A gene on chromosome 21 encodes APP. This gene contains 18 exons. The full-length protein contains 770 amino acids. It undergoes alternative splicing of exons 7, 8, and 15. The spliced protein missing exons 7 and 8 is 695 amino acids in length and is the predominant APP produced in the brain.

Amyloid precursor protein cleavage

APP is a transmembrane glycoprotein that is cleaved by secretase enzymes. Alpha-secretase is a metalloprotease of the ADAM family. Beta-secretase (BACE) is an aspartylprotease. Alpha- and beta-secretases cleave at different positions within the APP ectodomain (extracellular domain). APP is usually processed in the nonamyloidogenic pathway, where it is first cleaved by alpha-secretase to yield soluble APPs alpha and membrane-bound alpha-CTF (carboxyterminal fragment). Gamma-secretase then cleaves the membrane bound alpha-CTF to yield a small extracellular peptide P3 and an intracellular fragment AICD (APP intracellular domain).

In the amyloidogenic pathway, APP is first cleaved by beta-secretase to yield APPs beta, an extracellular soluble peptide, and membrane-bound beta-CTF. Gamma-secretase cleavage of the membrane-bound beta-CTF yields an extracellular component Abeta and intracellular AICD (Wilquet and De Strooper 2004; Figure 5–2).

The cleavage site of gamma secretase is heterogeneous, so that the Abeta peptides that are released differ in size. Most commonly, they contain 40 or 42 amino acids (Abeta 40, Abeta 42). Both of these are toxic. However, Abeta 42 is more insoluble and prone to aggregation. Abeta peptides stimulate oxidative stress. They may act as enzymes and produce H_2O_2 through reduction of metals and biological substrates (Newman et al. 2007). Diffusible oligomers of Abeta are present in cerebrospinal fluid in patients with AD. They consist of dimers, trimers, and tetramers derived from Abeta 40 and Abeta 42 species.

NONAMYLOIDOGENIC PATHWAY:

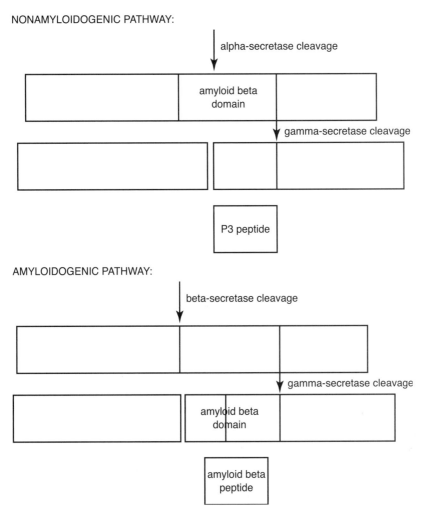

Figure 5–2. Cleavage of amyloid precursor protein by beta-secretase followed by gamma-secretase cleavage plays a key role in the generation of amyloid beta peptide. Excess production or deficient clearance of this peptide may lead to Alzheimer disease (based on Wilquet and De Strooper 2004; Selkoe 2002).

Function of APP

Amyloid precursor protein is a member of a family of proteins that includes amyloid precursor-like proteins 1 and 2. The biological functions of this protein family are not definitively known, though numerous functions have been proposed, including axonal transport, chaperone function, cell adhesion, and maintenance of metal balance.

Amyloid precursor protein APP695 is a membrane-anchored receptor for spondin, a secreted neuronal glycoprotein. Ho and Sudhof (2005) reported that the binding of F spondin inhibits cleavage of APP by beta-secretase. APP also interacts with a low-density lipoprotein (LDL). The soluble ectodomain-derived peptides generated from APP by secretase cleavage, APPsalpha and APPsbeta, have growth-promoting activities and increase proliferation of progenitor cells in the subventricular zone.

Amyloid and Pathophysiology of Alzheimer Disease

Role of soluble forms of amyloid

There is evidence from studies in transgenic mice that contain a mutant APP gene that synaptic terminals are depleted as the levels of soluble Abeta rise and before Abeta plaques occur. Selkoe (2002) noted that memory and cognitive deficits in humans also correlate better with levels of soluble cortical Abeta, including soluble oligomers.

Holscher et al. (2007) reported that the deleterious effects of the soluble amyloid fragments are reversible, indicating that treatment at the early stages of AD may prevent irreversible neuronal degeneration. These investigators reviewed evidence that soluble Abeta fragments impact synaptic plasticity in the hippocampus and impair memory formation. To investigate the effects of soluble Abeta fragments, they injected them into rat hippocampus and analyzed effects on synaptic function. Long-term potentiation was altered for several days following injection.

Role of amyloid aggregates

One hypothesis of AD postulates that the deposition and aggregation of misfolded proteins Abeta 42 are a key element in the disease. The most common cause of AD may, however, be a failure of degradation, so that the levels of Abeta progressively rise. Abeta 42 is initially present as low molecular weight oligomers that have subtle effects on synapses. Thereafter more complex aggregates form, often leading to diffuse plaque formation associated with inflammatory response and microglial activation (Selkoe 2002). In some forms of AD, Abeta accumulates primarily in brain parenchyma whereas in other forms it accumulates more extensively in cortical blood vessels, leading to cerebral amyloid antipathy (CAA). CAA is highly associated with the number of *ApoE4* alleles (Chalmers et al. 2003), discussed further in Figure 5–3 and on page 83.

Defective amyloid clearance

The finding that the Abeta-degrading protease neprilysin (NEP) is down-regulated in aging and in late-onset AD supports the concept that amyloid clearance is defective in AD. Farris et al. (2007) demonstrated through studies on transgenic mice that silencing the NEP gene led to prolonged half-life of Abeta dimers, increased the concentrations of Abeta dimers and the degree of hippocampal amyloid plaque, and promoted amyloid angiopathy. Farris et al. (2007) proposed that the reduction in concentration of NEP mRNA in aging and late-onset Alzheimer disease is due to oxidative damage of gene promoters or of proteins (Figure 5–3).

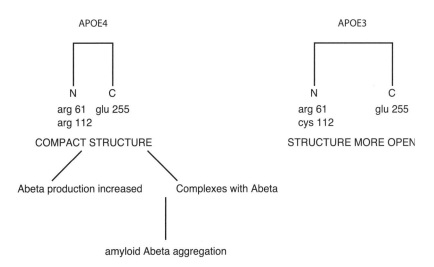

CNS TRAUMA

AGING

INCREASED OXIDATIVE STRESS

APOE4 GENOTYPE

DOMAIN INTERATION INTERACTION DIFFERS BETWEEN APOE4 AND APOE3

APOE4

N   C

arg 61   glu 255
arg 112

COMPACT STRUCTURE

APOE3

N   C

arg 61   glu 255
cys 112

STRUCTURE MORE OPEN

Abeta production increased   Complexes with Abeta

amyloid Abeta aggregation

Figure 5–3. Intense research efforts have led to the identification of important predisposing factors in the etiology of late-onset Alzheimer disease (based on Mahley and Huang 2006).

Tau in Alzheimer Disease

In addition to deposition of Abeta, the protein Tau is abnormally deposited in the brain in AD. The form of Tau that is deposited is hyperphosphorylated and polyubiquitinated. A less common form of dementia, frontotemporal dementia, is due to mutations in the Tau gene. Roberson et al. (2007) proposed that therapeutic measures be considered to reduce Tau deposition. Myers et al. (2005) reported evidence that individuals with Tau gene haplotypes that are associated with increased Tau protein production are at increased risk for AD.

Roberson et al. (2007) demonstrated that in transgenic mice expressing mutant human APP, genetic manipulations resulting in reduced endogenous Tau production prevented behavioral deficits. They noted that in the Tau-deficient mice beneficial effects were observed even though amyloid burden was not reduced. It is interesting to note that in AD the Tau that accumulates is abnormally phosphorylated and this affects its binding to microtubules. Microtubules play an important role in the transport of nutrient protein between the cell body and synapses.

Early-Onset Alzheimer Disease: Single Gene Defects

Autosomal dominant forms of AD are rare, representing 0.1% of cases. These forms may be due to mutations in the APP gene. Missense mutations in the APP-encoding gene were first described in patients with a rare familial form of AD. Missense mutations are often associated with impaired processing of APP. Frequently these mutations cluster around secretase cleavage sites.

Studies have revealed that the APP promoter mutations that lead to increased expression of APP are associated with AD. Mutations that result in increased production of Abeta due to duplication of the APP gene occur in rare forms of familial AD (Citron et al. 1994).

Gamma-secretase is a multimeric protein composed of subunits encoded by at least four different genes: presenilin PS1 and PS2, nicastrin, anterior pharynx defective Aph1 A or B, and presenilin enhancer. The presenilin/ gamma-secretase complex proteolytically cleaves protein other than APP. The gamma-secretase protein complex acts as a protease that cleaves transmembrane proteins within their transmembrane site. Gamma-secretase cleaves other proteins including the protein encoded by the Notch locus.

Presenilin variants occur and some of these are associated with early-onset AD. Most cases of this form of the disease are associated with presenilin 1 mutations and are inherited as autosomal dominant disorders (Taddei et al. 2002).

Late-Onset Alzheimer Disease

Linkage studies

In 1993 ApoE protein was found to be present in plaques in AD and to be bound to Abeta (Poirier et al. 1993). Pericak-Vance and Roses (1993) reported evidence for linkage of late-onset Alzheimer disease on chromosome 19q13. This chromosome region includes the *ApoE* gene. These studies represent the first indication that the apolipoprotein E locus is important in AD.

Genome-wide linkage studies on late-onset AD have yielded conflicting results. Differences in results likely reflect the fact that many genes and other factors contribute to etiology of this disease. Discrepant results may be due to admixture of individuals from different ethnic groups. Sillen et al. (2006) carried out studies in 71 Swedish families with AD in at least two generations. They noted that the population they studied is relatively genetically homogeneous. All affected individuals were typed for *ApoE4*. Their study included analysis of 365 genome-wide markers. Results in the whole sample set revealed significant linkage on chromosome 19q13. The maximum lod score was with a microsatellite marker located 3 Mb downstream of the *ApoE* locus. When *ApoE4*-positive families were evaluated, the maximum lod score was with a marker located 1.5 Mb upstream of *ApoE4*. These results support those of previous studies, indicating that there are possibly two different gene loci on chromosome 19q13 that determine AD. Sillen et al. (2006) noted that long haplotype blocks without recombination were seen in this region.

Association studies reveal importance of apolipoprotein E in Alzheimer disease

Coon et al. (2007) described results of a whole genome association study using 502,627 single nucleotide polymorphisms (SNP) in 1086 histologically verified cases of AD and controls. The SNP analysis was performed using the 500K Affymetrix chip. The patient samples included 362 females and 302 males, mean age 82 ± 7.7 years. The mean age of the 422 controls was 79 ± 11 years. They noted that SNP genotyping precisely identified the *APOE* region as having significant association with late-onset AD. No SNPs are present within the *APOE* gene. The *APOE* region was far more strongly associated with AD than any other of the 502,627 tested. The significance of the association with an SNP 14 kb from the *APOE* gene was 1×10^{-39}. These studies indicate that the odds ratio of the overall risk of AD in *APOE4* homozygotes is 25.3 times greater than that in *APOE3* homozygotes.

All samples were also genotyped for *APOE* gene polymorphism. The significance of the association with the *APOE4* allele was 1×10^{-44}. The study of cases and controls subjected to histological studies at autopsy is significant given the fact that 10% of clinical diagnosed cases of AD do not have neuropathological criteria for the disease and one-third of elderly patients not diagnosed with AD have neuropathological evidence of the disease (Reiman 2007).

Additional Evidence for Genetic Factors: Family History and Twin Studies

First-degree relatives of a patient with AD have a 2.5 times increased risk of AD. If the background lifetime risk is 10% (the most common estimate used), the risk of AD in a first-degree relative of a patient is 25% (Silverman et al. 1994). There is a major association between family history and the risk of developing AD even beyond the *ApoE* genotype (Reiman et al. 2007) and known autosomal dominant mutations. Family risk not associated with *APOE* genotype may be attributable to as yet unidentified genes.

Gatz et al. (2006) carried out an analysis of 11,884 twin pairs born prior to 1935. The initial screen was for cognitive dysfunction. They used specific diagnostic criteria to distinguish between cases with vascular dementia and AD, and to distinguish cases that may have mixed forms of vascular dementia and AD, or other forms of dementia. Based on the use of different models adjusting for age, they estimated heritability, that is, the percentage of risk due to genetic factors. Heritability estimates varied between 58% and 79%, and intrapair differences were greater for dizygotic (DZ) twin pairs.

Although concordance for AD was higher in monozygotic (MZ) twins than in DZ twins, concordance for vascular dementia did not differ between DZ and MZ twins. The average age of onset of AZ disease was 78.1 years in MZ twins and DZ twins. In 25 pairs of MZ twins where both had AD, the difference in age of onset was 3.66 ± 3.63 years. Among AD-concordant DZ twin pairs, the difference in age of onset of symptoms was 8.12 ± 7 years. These findings indicate that genes play a role in the age of onset of symptoms. The authors emphasize that though the concordance for AD in MZ twins is not perfect, it is very high, 0.81 in males and 0.89 in females. The concordance rate for AD in like-sex male DZ twins was 0.48. In like-sex female DZ twins, AD concordance was 0.76. In unlike-sex DZ twins, concordance rate for AD was 0.58.

APOE AND *APOE4* VARIANT AS RISK FACTORS
FOR DEMENTIA

ApoE is a 299 amino acid protein encoded by a gene on chromosome 19q13. The ApoE-encoded protein is abundant in brain. Apolipoproteins are carrier proteins for lipids and ApoE is the protein carrier for very low density lipoprotein. It is also a carrier for cholesterol and fat-soluble vitamins. Lipoproteins contain a central core of nonpolar lipids (triglycerides and cholesterol esters) and a surface monolayer of polar lipids and apoproteins. The apoproteins increase the water solubility of the lipoprotein particles.

As described above, early indications of the importance of *APOE* came from studies that revealed the linkage of AD to chromosome 19q13. *ApoE* is located in this chromosomal region. Further evidence derives from the fact that *APOE* colocalizes with amyloid and there is evidence that the different *APOE* isoforms interact differently with Abeta.

APOE synthesis occurs primarily in the liver and in the brain. Glial cells that surround sensory and motor neurons and nonmyelinating Schwann cells produce APOE. In the central nervous system, APOE is produced by astrocytes. APOE mRNA also occurs in neurons in the cortex and hippocampus. Mahley et al. (2006) have presented evidence that the neuronal expression of *ApoE* promotes neuron protection and repair. The N terminal domain, amino acids 1–191 of *APOE*, contains an LDL receptor-binding site. The C terminal domain, amino acids 216–299, contains a lipid-binding site. APOE interacts with LDL receptors to distribute lipids throughout the central nervous system. Lipids including cholesterol are distributed to the site of nerve injury.

Genetic Variation

There are three common allelic variants of this gene. The most common allele, designated *E3*, occurs with a frequency in the U.S. population of 60% to 70%. The *APOE4* allele occurs with a frequency of 15% to 20%, and the *APOE2* allele occurs in 5% to 10% of the population. The proteins derived from these alleles differ at two sites, aminoacid (amino acids) 112 and 158:

> *APOE3* 112 cys (cysteine) 158 Arg (arginine)
> *APOE4* 112 arg 158 arg
> *APOE2* 112 cys 158 cys

Cys at position 158 affects the binding of apolipoprotein E to its receptor and predisposes to a form of hyperlipoproteinemia type III. This allelic variation does not inevitably lead to hyperlipoproteinemia. This allelic variant interacts with other genetic and environmental factors.

The occurrence of arg at position 112 in *APOE4* impacts the protein structure. Arg112 changes the folding and molecular conformation such that arg61 extends outward and is able to interact with Glu 255 in the C terminal region. This interaction alters a number of properties of *APOE4*, including protein stability. *APOE* is cleaved by proteases in neurons. The structural changes in *APOE4* make it more susceptible to protein cleavage than *APOE3* (Figure 5–4). The *APOE4* peptide containing amino acids 1–272 is toxic in cultured neurons and leads to neurodegenerative changes. These toxic fragments interact with the neuronal cytoskeleton and with mitochondria, and may induce cell death.

There is also evidence that apolipoprotein has an important effect on the deposition and clearance of Abeta protein. The carboxyterminal fragment of *APOE* copurifies with Abeta peptide found in senile plaques.

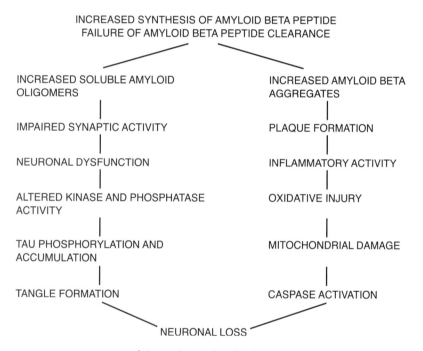

Figure 5–4. Diagram of the pathway that leads to neuronal loss in late-onset Alzheimer disease (based on Selkoe 2002; Mahley et al. 2006).

ApoE and Risk of Alzheimer Disease

Farrer et al. (1997) reported that the risk of developing AD is four to nine times greater for individuals who are homozygous for the *ApoE4E4* homozygotes than for *ApoE3E3* homozygotes. The risk for *E3E4* heterozygotes was 3.2 times that in *E33* homozygotes. The *ApoE4* allele also results in increased risk of AD and is also increased in Chinese and Japanese populations and in African Americans. It is important to note, however, that one-third of patients with AD lack the *ApoE4* allele. Furthermore, 50% of *E4* homozygotes survive to 80 years without developing AD.

Biology of ApoE in Alzheimer Disease

Ye et al. (2005) reported that Abeta production is related to cellular cholesterol concentration. The question then arises whether *ApoE* influences amyloid through its lipid-binding capacity. There is evidence that Abeta synthesis is increased in the presence of *ApoE4* alleles (Vance et al. 2006). Individuals with the *ApoE4* allele have greater amyloid deposition and tangle pathology and worse mitochondrial damage than patients without this allele.

Mahley and Huang (2006) proposed that *ApoE* plays a critical role in AD neurodegeneration and that the structural features of *ApoE4* are particularly important in this regard. They demonstrated that the amino acid substitutions in *ApoE4* cause the N terminal and C terminal domains of *ApoE4* to interact. They demonstrated that neurons synthesize ApoE. In neuronal cell cultures, they determined that *ApoE4* potentiates lysosomal leakage and apoptosis that is induced by Abeta. Furthermore, *ApoE4* is more susceptible to proteolytic cleavage than the other forms of ApoE and that these products of *ApoE4* lead to mitochondrial degeneration. They therefore proposed therapeutic strategies with small molecules that impact *APOE4* structural interaction and the use of protease inhibitors that prevent generation of toxic *ApoE4* fragments. Studies are also underway to identify molecules that promote clearing of proteolytic cleavage products of *ApoE4*.

LIPID METABOLISM AS RISK FACTOR: CHOLESTEROL AND LIPIDS' RELEVANCE TO DEMENTIA

Cholesterol is synthesized in the brain, and cholesterol does not cross the blood-brain barrier; therefore the brain cholesterol level is independent of the blood cholesterol level. Cholesterol is degraded in the brain by the enzyme 24-hydrocholesterol dehydrogenase and by oxidation by cytochrome CYP46 (Canevari and Clark 2007). Cholesterol is abundant in myelin and in neuronal cell membranes. Cholesterol and ApoE are synthesized in glial cells. The transporter ABCA facilitates secretion from these cells.

Small amounts of cholesterol are synthesized by neurons early in development. ApoE and cholesterol are necessary for synapse formation. Nuclear factors LXR (liver X receptor) are abundant in brain and as transcription factors play an important role in cholesterol homeostasis. Oxysterols and other agonists of LXR increase ABCA1 expression and cholesterol efflux. Cholesterol affects APP processing. It impacts the activities of beta- and gamma-secretase and the compartmentalization of these enzymes. There is evidence that efflux of cholesterol from neurons and astrocytes is less efficient in individuals with the *ApoE4* genotype. The presence of *ApoE4* in lipoproteins leads to enhanced deposition of Abeta.

ENVIRONMENTAL RISK FACTORS: NEUROTOXIC CHEMICALS, ANESTHESIA

There are a number of reports that indicate that perioperative factors increase neuropathology in AD. These factors include anesthetics, hypocapnia, and hypoxia and may lead to marked postoperative cognitive dysfunction. Eckenhoff (2004) demonstrated that isofluorane enhanced aggregation of Abeta.

Xie et al. (2007) demonstrated that at concentrations currently used for anesthesia, isofluorane alters amyloid. Xie et al. reported that isofluorane induces apoptosis in cultured neuroglioma cells. This in turn increases levels of the enzyme beta site APP-cleaving enzyme BACE and gamma-secretase and secretion of Abeta. They noted further that isofluorane increases Abeta aggregation (Figure 5–5).

Postoperative delirium occurs in 15% to 53% of surgical patients following general anesthesia. There is evidence that this may contribute to postoperative cognitive dysfunction. This is particularly well documented following coronary bypass surgery. There is evidence that the age of onset of AD is inversely correlated with the cumulative exposure to anesthesia before age 50. Xie et al. (2007) noted that there is increasing evidence that dementia is a leading risk factor for postoperative delirium. Since isofluorane induces neuroglioma cell apoptosis and enhances Abeta oligomerization, there is likely a link between postoperative delirium and long-term dementia. They noted that studies to

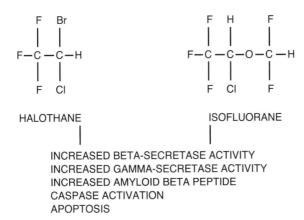

Figure 5–5. Molecular structures of volatile anesthetics halothane and isofluorane. Exposure to these agents increases the risk of Alzheimer disease. Isofluorane induces apoptosis of neuroglioma cells in culture and increases production of amyloid beta peptide (Xie et al. 2007).

examine the epidemiologic association of isofluorane anesthesia and AD are urgently needed.

Studies of Eckenhoff and colleagues (2004) revealed that both halothane and isofluorane significantly accelerated oligomerization of Abeta 42. They demonstrated further that in cells transfected with a presenilin mutation there was greatly increased sensitivity to isofluorane toxicity. Note that presenilin is a determinant of gamma-secretase function. Eckenhoff et al. noted that the most commonly used inhaled anesthetics are small haloalkanes or haloethers that bind to hydrophobic cavities in proteins. They also bind to interpeptide interfaces and stabilize oligomers.

There is also direct evidence in animal studies that isofluorane induces apoptotic neurodegeneration (Perouansky 2006). Studies on aged rats revealed that inhalation anesthetics caused lasting impairment of spatial memory performance.

Mandal et al. (2006) reported that multidimensional nuclear magnetic resonance (NMR) spectroscopy revealed that halothane specifically interacts with Abeta 40 and Abeta 42. Halothane changes the Abeta peptide from an alpha-monomeric alpha-helical form to an oligomeric beta-sheet conformation. On the basis of NMR studies, Mandal et al. (2007) reported that the oligomerization propensity of Abeta was highest for halothane and isofluorane and slightly less for propofol (Figure 5–6).

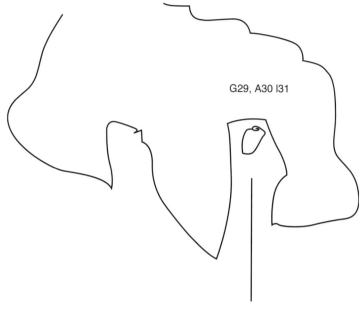

G29, A30 I31

HALOTHANE INTERACTS WITH AMIDE
PROTONS IN AMINO ACIDS G29, A30, I31
INAMYLOID BETA PEPTIDE

Figure 5–6. Halothane and isofluorane react with specific amide protons within the amyloid beta peptide and convert the alpha-helix structure to a beta pleated sheet structure; this increases oligomerization (Mandal et al. 2006).

TRANSLATIONAL ASPECTS AND THERAPY

Gamma-Secretase Inhibitors

Walsh and Selkoe (2004) presented evidence that Abeta oligomers may interfere with synaptic function. They noted, however, that substantial or complete inhibition of production of Abeta monomers would likely be deleterious. They noted that administration of gamma-secretase inhibitors would therefore need to be titrated to suppress Abeta production to levels that do not support oligomerization but not cause deficiency of monomers.

Gamma-secretase inhibitors are in clinical trials in AD. Siemers et al. (2006) reported that a gamma-secretase inhibitor administered to AD patients for 6 weeks resulted in significant decreases in plasma levels of Abeta, but no decrease in cerebrospinal fluid Abeta levels were observed.

Inhibitors of Amyloid Beta Aggregation

Based on evidence that AD is due to accumulation of neurotoxic oligomeric aggregates, McLaurin et al. (2006) carried out studies to examine in transgenic mice the effects of inhibitors of Abeta aggregation on the pathogenesis of AD. They reported that oral administration of the molecule cyclohexanehexol and its stereoisomers block Abeta oligomer accumulation and the consequent pathological and neurobehavioral effects. Their choice of cyclohexanehexol was based on their previous studies that demonstrated that phosphatidylinositol lipids facilitate Abeta oligomerization and fibril formation. They postulated that derivatives of phosphatidylinositol may compete for binding to Abeta. McLaurin et al. initially demonstrated in an in vitro system that cyclohexanehexol stereoisomers could inhibit Abeta fibril assembly and act to stabilize Abeta as nontoxic conformers. They demonstrated that different stereoisomers of cyclohexanehexol have different effects.

To investigate the efficacy of treatment at different stages of AD, McLaurin et al. carried out experiments in a mouse model of human AD. They assigned littermates to different treatment paradigms: prophylactic treatment or treatment after the onset of symptoms. The study endpoints were cognitive function measured by spatial reference learning, brain Abeta levels, neuropathology, and mortality. Their studies demonstrated that prophylactic treatment significantly delayed the onset of disease symptoms. Results demonstrated that the cyclohexanehexol stereoisomer significantly inhibits AD behavioral deficits and neuropathology. McLaurin et al. postulated that these compounds are potentially suitable for treatment of AD in humans. They are transported into the central nervous system by facilitated mechanisms and have high bioavailability there. Furthermore, they are metabolized to glucose in the central nervous system.

Potential Modulation of Lipid Metabolism to Treat Alzheimer Disease

Apolipoproteins are carrier proteins for lipids and ApoE is the protein carrier for very low density lipoprotein. It is also a carrier for cholesterol and fat-soluble vitamins. ApoE colocalizes with amyloid and there is evidence that the different ApoE isoforms interact differently with Abeta.

Ye et al. (2005) reported that Abeta production is related to cellular cholesterol concentration. The question then arises whether ApoE influences amyloid through its lipid-binding capacity. ApoE and lipoproteins are synthesized in glia and then deliver cholesterol to neurons for axon repair and growth. There is evidence that Abeta synthesis is increased in the presence of *ApoE4* alleles (Vance et al. 2006). There is evidence that the binding of amyloid is lipid type dependent and ApoE isoform dependent (Hirsch-Reinshagen and Wellington 2007).

Cellular uptake of Abeta is dependent on receptors, neurons, and activated astrocytes expressing lipoprotein receptor related protein. The LDL receptor does not apparently play a role in amyloid metabolism.

Statins in treatment of Alzheimer disease

A number of studies have been undertaken to determine if statins impact the course of AD. The results are, however, not clear-cut. Zhou and coworkers (2007) carried out a meta-analysis of data published on AD treatment with statins. They determined that there was no beneficial effect. Riekse et al. (2006) reported that the results of treatment of AD with HMG CoA reductase inhibitors (statins) may vary depending on their ability to cross the blood-brain barrier and penetrate the brain.

Promoters of cholesterol and Abeta efflux: agonists of LXR

Bell et al. (2007) published studies on the clearance of amyloid beta peptide from the nervous system across the blood-brain barrier. They demonstrated that lipoprotein receptor–related proteins LRP1 and LRP2 play important roles in the clearance of Abeta and its transport across the blood-brain barrier. They demonstrated the Abeta 42 is cleared half as rapidly as Abeta 40. Clearance of Abeta complexed to ApoE was slowed; however, Abeta complexed to glial cell–derived apolipoprotein J was more rapid. Apolipoprotein J is a major carrier of Abeta.

There is evidence that low intracellular cholesterol levels reduce Abeta production while high levels increase Abeta production. It is important to consider the process of efflux of cholesterol across cell membranes. The ATP-binding cassette protein ABCA1 plays a role in the efflux of excess cellular cholesterol, and its loading onto lipid-poor apolipoproteins. ABCA1 deficiency in mice leads to a decrease in ApoE levels in mouse brain ApoE is poorly lipidated; however, amyloid deposition is decreased.

Liver X receptors are transcription factors that enhance transcription of a number of genes, including genes involved in cholesterol efflux. Molecules that act as LXR agonists, such as oxysterols, therefore promote cholesterol efflux and reduce amyloid generation in mouse models of AD.

Small Molecules That Impact *APOE4* Structure and Proteolysis

Mahley and Huang (2006) proposed that *ApoE* plays a critical role in AD neurodegeneration and that the structural features of *ApoE4* are particularly important in this regard. They demonstrated that the amino acid substitutions in *ApoE4* cause the N terminal and C terminal domains of *ApoE4* to interact. They demonstrated that neurons synthesize ApoE. In neuronal cell cultures, they determined that *ApoE4* potentiates lysosomal leakage and apoptosis that is induced by Abeta. Furthermore, *ApoE4* is more susceptible to proteolytic cleavage than the other forms of *ApoE* and these products of *ApoE4* lead to mitochondrial degeneration. They therefore proposed therapeutic strategies with small molecules that impact *APOE4* structural interaction and the use of protease inhibitors that prevent generation of toxic *ApoE4* fragments. Studies are also underway to identify molecules that promote clearing of proteolytic cleavage products of *ApoE4*.

EMPIRIC OBSERVATIONS ON BENEFICIAL TREATMENT OF ALZHEIMER DISEASE

Dissection of the molecular basis of empiric observations may lead to development of more efficacious treatments.

Nonsteroidal Anti-inflammatory Drugs

Epidemiologic studies on AD revealed that the risk of disease is apparently reduced in individuals who regularly use nonsteroidal anti-inflammatory drugs (NSAIDs). Eriksen et al. (2003) reported that molecular studies indicate that a 30% increase in Abeta 42 leads to symptom development. They noted that in addition to Abeta deposition in plaques and blood vessel walls, together with development of neurofibrillary tangles and neuronal loss, there is often a marked inflammatory response in the vicinity of the Abeta deposition. They postulated that the inflammatory response may be harmful and promote further Abeta precipitation and that anti-inflammatory drugs may be helpful. Eriksen et al. (2003) examined the effects of a number of different NSAIDs on levels of Abeta in the H4 neuroglioma cell line that expresses the APP695NL mutation of Abeta peptide. Their studies in cultured cells and in mouse brain revealed that some but not all NSAIDs reduced Abeta 42. They determined that two commonly used NSAIDs, ibuprofen and indomethacin, significantly lower Abeta 42 levels in APP mice. Naproxen had no effect.

In an effort to identify NSAIDs with reduced gastrointestinal and renal toxicity, they explored the action of ibuprofen enantiomers. They determined that flurbiprofen and sulindac are equally effective in reducing Abeta in mouse models of AD and can be used in higher doses than ibuprofen. These drugs are now in clinical trials in humans. The studies of Eriksen et al. (2003) also demonstrated that flurbiprofen reduces Abeta production by targeting gamma-secretase.

Antioxidants

Given the important role of ROS and superoxides in inducing oxidative DNA damage, there are active investigations to identify natural antioxidants. Kawanishi et al. (2005) emphasized that extensive safety studies of these compounds are required since some antioxidants are metabolized in the body to yield pro-oxidants. One such compound is phytic acid.

Physical and Mental Activity

There is epidemiologic data that supports the view that physical and mental exercise and social engagement delay the onset of AD (Marx 2004).

6

ADVANCES IN TREATMENT OF SINGLE GENE DISORDERS THROUGH TRANSLATIONAL SCIENCE: TREATMENT WITH SMALL MOLECULES (CHAPERONES) AND GENE-BASED THERAPIES

It is likely that the deleterious effects of protein misfolding are relevant to the pathogenesis of a number of genetically determined diseases. Gregersen (2006) postulated that pharmacological chaperones might be of value in treatment of a large array of genetic diseases. Chaperones have proven useful in treatment in some cases of lysosomal storage diseases.

Physiological chaperones occur in many organisms, including humans; they play a role in protein folding. One purpose of such folding is to shield hydrophobic domains in polypeptide chains and proteins to protect these residues from interaction with other polypeptides. Heat shock proteins represent a class of chaperone proteins produced in response to cellular stresses, including heat and oxidative stress. The binding of physiological chaperones to hydrophobic domains in polypeptides is energy dependent; adenosine triphosphate (ATP) is hydrolyzed to ATP in the reaction. Chaperones are also released from folded proteins.

Amino acid substitutions in proteins may impair folding. High concentrations of misfolded proteins may overwhelm the cellular systems to degrade these proteins. Misfolded proteins are eliminated by intracellular proteases and by the ubiquitin proteosome system. The cell's ability to cope with misfolded proteins deteriorates with aging. This in turn may promote the formation of protein aggregates. This mechanism may play a role in the accumulation of alpha-synuclein aggregates in Parkinson disease and amyloid beta aggregates in Alzheimer disease.

Pharmacological chaperones are actively being developed and investigate to determine their effectiveness in promoting small molecule stabilization (Gregersen 2006). In cases where gene mutations or deletions cause protein misfolding but do not significantly impair functional domains, chaperones may serve to facilitate stabilization of the misfolded protein and increase the concentration of functional protein.

LYSOSOMAL ENZYME DEFICIENCIES: TREATMENT WITH SMALL MOLECULES (CHAPERONE THERAPY)

Lysosomal storage diseases are due to mutations in enzymes that function within lysosomes to degrade complex glycoproteins, glycolipids, and glycosphingolipids. These diseases may also result from deficiency of enzymes and proteins involved in lysosomal enzyme posttranslational modification and organelle targeting. Disease manifestations result from accumulation of undegraded substrates, and manifestations may occur in infancy, in childhood, or only during adult life, depending on the degree of impairment of enzyme function. Multiple organ systems are often involved; organs most commonly affected include liver, spleen, bone, brain, kidney, and muscle (Neufeld 1991).

Treatment With Small Molecules

Lysosomal enzymes exhibit maximal kinetic activity at low pH. Small molecules that are enzyme inhibitors in in vitro assays at low pH may act to stabilize mutant enzymes in the cytoplasm at neutral pH. This stabilization facilitates transport of the mutant enzyme from the cytoplasm to the lysosome. The low pH that exists within the lysosome displaces the inhibitor from the mutant enzyme. Mutant enzymes that have an intact active site may then be active. Fan et al. (1999) first demonstrated that a small molecule inhibitor of alpha-galactosidase was able to increase the activity of mutant alpha-galactosidase in Fabry disease. Specific mutations of alpha-galactosidase respond to therapy with 1-deoxygalactonojirimycin.

In a number of different lysosomal diseases the causative gene mutations, such as small deletions or missense mutations, often lead to protein misfolding. In cases where these mutations do not significantly alter functional domains, chaperones may facilitate stabilization of the misfolded protein. Analogues of lysosomal enzyme substrates, particularly iminosugars, may facilitate stabilization of misfolded protein. The most frequently used in therapy of these disorders are deoxynojirimycin analogues (Beck 2007). The effect of chaperones varies depending on the nature of the underlying mutation. For example, in the case of one of the beta-glucocerebrosidase mutations that lead to Gaucher disease, N370S, activity of the enzyme protein is increased by N-nonyl-deoxynojirimycin, NN-DNJ). However, NN-DNJ does not increase activity of the L444P mutant enzyme. Interestingly, 1-deoxygalactonojirimycin increases activity of the wild-type beta-glucosidase enzyme. Activity of both the N370S and the L444P mutant proteins in Gaucher disease is increased by an iminosugar chaperone, isofagimine (Brooks 2007; Figure 6–1).

Hexosaminidase mutations present in adult-onset forms of the GM2 gangliosidosis, Tay-Sachs disease, and Sandhoff disease do not affect the active site of the enzyme. In vitro studies have demonstrated that activity of hexosaminidase in these cases can be enhanced by chaperone therapy N-acetylglucosamine thiazoline. GM1 gangliosidosis responds to therapy with N-octyl-4-epi-beta valienamine (Matsuda et al. 2003).

Pompe disease results from mutations in the alpha-glucosidase gene. This disorder is inherited as an autosomal recessive and is characterized by excessive glycogen accumulation in skeletal muscle and in cardiac muscle. Severely affected infants succumb to cardiac and respiratory failure. Enzyme therapy (Myozyme) is approved for therapy (see p. 99); however, because of cost and because of immune response to infused enzymes, chaperone therapy is an important consideration. Kakavanos et al. (2006) reported that D-glucose is a competitive inhibitor of alpha-glucosidase. D-Glucose also serves to prevent aggregation and precipitation of mutant protein that results from several different specific missense mutations in the alpha-glucosidase encoding gene.

Substrate Reduction Therapy

Radin (1996) proposed that reduction of substrates for lysosomal enzymes could be achieved through inhibition of enzymes involved in glycosphingolipid synthesis. Iminosugars are known inhibitors of alpha-glucosidase. Cox et al. (2000) reported results of treatment of 28 adult patients with Gaucher disease with N-butyldeoxynojirimycin and noted that liver and spleen enlargement was significantly reduced by treatment. In 2003, an international advisory council established criteria for use of N-butyldeoxynojirimycin in

GAUCHER DISEASE; EFFECT OF MUTATIONS
IMPACT OF CHAPERONE THERAPY

MOST COMMMON MUTATION: N370S, L444P

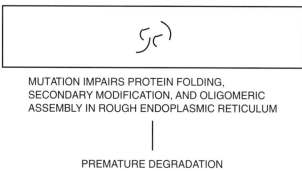

MUTATION IMPAIRS PROTEIN FOLDING,
SECONDARY MODIFICATION, AND OLIGOMERIC
ASSEMBLY IN ROUGH ENDOPLASMIC RETICULUM

PREMATURE DEGRADATION

EFFECT OF CHAPERONE IMINOSUGAR ISOFAGIMINE

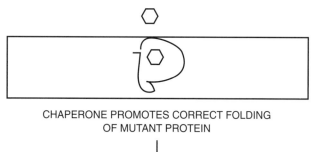

CHAPERONE PROMOTES CORRECT FOLDING
OF MUTANT PROTEIN

PREVENTS DEGRADATION
INCREASES TRAFFICKING OUT OF
ROUGH ENDOPLASMIC RETICULUM

Figure 6–1. An imino-based molecular chaperone restores folding and trafficking of specific mutants of beta-glucosidase proteins and constitutes an effective treatment in Gaucher disease (based on Brooks 2007).

treatment of type 1 Gaucher disease. *N*-butyldeoxynojirimycin is also a treatment option for GM1 and GM2 gangliosiosis and for the adult form of Tay-Sachs disease and for the adult form of Niemann-Pick type C disease.

Iminosugars therefore act not only as enzyme inhibitors to achieve substrate reduction but also as chaperones.

Enzyme Replacement Therapy

Enzyme replacement therapy (ERT) is in clinical use or in clinical trials for treatment of a number of different lysosomal storage diseases. Problems related to ERT include, first, the extraordinarily high cost of treatment, the minimal effects on storage in brain and bone, and the risk of induction of immunological response that may reduce or neutralize activity of infused enzymes.

Beck (2007) drew attention to the difficulties inherent in establishing study protocols for lysosomal storage disease therapies. These include the marked heterogeneity that exists between patients in clinical manifestations and rates of progression and the relatively small numbers of patients available for study. In a number of these disorders, surrogate markers rather than clinical symptoms are monitored to ascertain treatment efficacy. For example, in Fabry disease the degree of clearance of stored substances, specifically microvascular endothelial deposits, was analyzed. In mucopolysaccharide storage disease, urinary excretion of glycosaminoglycans is monitored. More recently, heparin cofactor II thrombin complex was shown to be an effective indicator of storage and treatment efficacy. In Pompe disease, urinary concentrations of a tetrasaccharide, Glu4, are monitored. This tetrasaccharide arises through intravascular degradation of glycogen. In type 1 Gaucher disease, sizes of liver and spleen are effectively reduced by enzyme replacement therapy, and spleen and liver size reduction serves as a biomarker. However, skeletal manifestations in Gaucher disease respond slowly. The neuronopathic form of Gaucher disease does not respond to enzyme replacement therapy.

Enzyme replacement therapy is used for treatment of Gaucher disease, Fabry disease, mucopolysaccharidoses including MPS1 (Hurler disease) and MPS2 (Hunter disease), Morquio disease, Maroteaux-Lamy disease, glycogen storage disease, and Pompe disease. In reviewing studies on immunological response to enzyme replacement therapy, Beck (2007) concluded that ERT is generally well tolerated. Patients who develop antibodies usually continue to tolerate enzymes, though they may require a higher dosage.

In developing preparations for ERT, knowledge about the cellular uptake of enzyme preparations is essential. For a number of lysosomal enzymes, tagging with mannose-6-phosphate is essential. Other lysosomal enzymes utilize different receptors; these include a mannose receptor and an asialoglycoprotein receptor. The mannose receptor is essential for uptake of glucosylceramide (beta-glucosidase) into macrophages. Successful treatment of Gaucher disease requires that enzymes produced through recombinant methodology expose mannose groups (Van Patten et al. 2007).

Modifiers and Epigenetic Factors: Impact on Treatment Response

There is a growing body of data that indicates that patients with the same disease who have the same underlying mutation may not have the same clinical course or the same response to treatment. Goker-Alpan et al. (2005) reviewed cases of Gaucher disease with a specific beta-glucosidase mutation L444P. Patients homozygous for this mutation are considered to have the neuronopathic form of Gaucher disease. Based on phenotypic features, however, 25% of L444P homozygous patients were classified as Gaucher disease type 1. Their review of clinical records of 70 patients revealed marked variation in clinical course. Developmental outcome ranged from normal cognition to developmental delay with IQ less than 69. Three of the patients were diagnosed as having autism. They noted that while all patients in the cohort studied had neurological symptoms, the severity of these varied.

Enzyme levels among the 70 patients varied from 1% to 13.3% of normal. Importantly, five cases where enzyme levels were at the higher end of the spectrum had severe disease. These findings have implications for treatment of the disorder. Goker-Alpan et al. emphasized that their findings indicate that modifier genes play an important role in the disease phenotype. Goker-Alpan et al. noted that gene modifiers may impact the intracellular processing of the enzyme or the metabolism of beta-glucosidase, or glucocerebroside.

It is interesting to note that there is a beta-glucosidase pseudogene that occurs 16 kb downstream of the functional gene. Rearrangements between the expressed gene and the downstream gene sometimes occur. Perhaps better understanding of the clinical heterogeneity may evolve from comprehensive gene analysis.

ADVANCES IN TREATMENT OF PHENYLKETONURIA USING COFACTOR THAT ACTS AS A CHAPERONE

Early detection of phenylketonuria (PKU) through determination of blood phenylalanine levels in the newborn period, followed by treatment with a low-phenylalanine diet, are effective in preventing the mental retardation that is characteristic of untreated PKU (Abadie et al. 2001). However, long-term treatment with the highly restrictive diet is often difficult.

Mutations in the phenylalanine hydroxylase encoding gene are the major cause of raised blood phenylalanine levels. Rare causes of raised phenylalanine levels include defects in the synthesis and metabolism of tetrahydrobiopterin, which acts as a cofactor for phenylalanine hydroxylase. In 2004, Kure et al. demonstrated that individuals with specific phenylalanine hydroxylase mutations respond favorably to treatment with tetrahydrobiopterin and several of its stereoisomers. There is evidence that tetrahydrobiopterin acts as a chemical chaperone that prevents misfolding of mutant phenylalanine hydroxylase. Mutant forms of this enzyme that are misfolded may be rapidly degraded through ubiquitin-mediated proteolysis. A number of different phenylalanine hydroxylase mutations are tetrahydrobiopterin responsive. Current recommendations are that responsiveness to this compound be determined in all patients with raised phenylalanine levels. In responsive patients, tetrahydrobiopterin treatment permits dietary liberalization.

MOLECULAR METHODS TO MODIFY GENE EXPRESSION; SHORT OLIGONUCLEOTIDES AND RNA INHIBITION

Availability of human genome sequence information has stimulated the search for methods to specifically silence gene expression. Methods that have emerged involve RNA inhibition (RNAi) and use of short oligonucleotides.

Antisense Oligonucleotides

Antisense oligonucleotides bind to unprocessed messenger RNA (mRNA) and may affect the splicing. Antisense oligonucleotides can be used to induce skipping of exons that contain deleterious mutations. Antisense oligonucleotides are in use in clinical trials to treat some forms of Duchenne muscular dystrophy. Exon skipping may result in production of a truncated but functional dystrophin protein. Modification of antisense oligonucleotides with morpholino-related chemicals delays their degradation and permits intravenous injection of these compounds.

Allele-specific oligonucleotides are used to silence expression by binding to mRNA and then inducing degradation by RNA H (Opalinska and Gewirtz 2002). Binding of an allele-specific oligonucleotide may also induce steric hindrance of the ribosomal machinery that impairs translation of mRNA to protein.

RNA Inhibition

Silencing of gene expression by double-stranded RNA was first observed in plants, and Fire et al. subsequently analyzed details of this process in

Caenorhabditis elegans in 1998. They determined that double-stranded RNA significantly interferes with gene expression and that this constitutes a physiological gene-silencing mechanism. The enzyme Dicer cleaves double-stranded RNA to yield short inhibitory RNA (siRNA). These are then incorporated into a specific complex, the RNA-induced silencing complex (RISC). This complex causes the double-stranded siRNA to unwind. The sense strand of the siRNA is degraded by nucleases, and the antisense strand binds to the target mRNA sequence. Following this annealing, the target mRNA sequence is then degraded through the endonuclease activity in the RISC.

siRNA technology has become increasingly important for in vitro studies to analyze the physiological effects of gene knockdown. With respect to the therapeutic applications of siRNA technology, it is important to emphasize that problems exist with respect to both cellular delivery of siRNA and plasma stability. siRNA has been used locally through injection into the eye to treat the wet or vascular form of age-related macular degeneration.

Problems Related to RNAi Therapy

There is evidence that in some instances a specific siRNA may bind to target sequences with partial homology. Wall and Shi (2003) noted that stringent RNAi design is required to avoid off-target activity.

In mammals, siRNA gene activation is usually transient. Intracellular introduction of plasmids that produce short hairpin RNA can circumvent this problem. However, given the problems related to viral and vector-mediated gene therapy, other methods are being pursued to enable RNAi introduction into cells.

Intranuclear Transcriptional Silencing

RNA inhibition also takes place in the nucleus. siRNA complementary to $5'$ gene regions leads to transcriptional silencing in a not clearly defined mechanism that includes chromatin modification and histone methylation (Kim et al. 2007).

Micro-RNA

Mammalian genomically encoded micro-RNAs are transcribed by polymerase II and processed by a specific complex, Drosha-DGCR8 (DGCR8 is encoded by the Di George critical region 8 on chromosome 22). The pre-micro-RNAs are often 70 nucleotides in length and they occur as double-stranded hairpin structures that exhibit self-complementary binding. They are transported to the cytoplasm through the activity of exportin. There they are cleaved by the Dicer enzyme and then react in the RISC-AGO (endonuclease protein) complex. Micro-RNAs target sequences in the 3' untranslated regions of mRNA and lead to inhibition of translation and mRNA degradation. There is evidence that micro-RNAs play an important role in control of gene expression during development and differentiation (Kim and Rossi 2007).

GENE THERAPY

On Hold Awaiting Better Delivery Systems

Gene therapy trials with adenovirus vectors were put on hold in 1999 following the death of a patient with ornithine transcarbamylase (OTC) deficiency that was treated with adenoviral-OTC gene recombinant virus. Adenoviral vectors often elicit a marked immune response. Retroviral vectors are often preferred in gene therapy because they are likely to be more stably integrated. However, the fact that such vectors led to tumor formation in three boys treated for X-linked severe combined immunodeficiency has raised concern about their use (Couzin and Kaiser 2005). Currently, adeno-associated viral vectors are favored for use in gene therapy since they have the potential for long-term expression and are less likely to induce inflammatory and antibody responses.

Hemophilia: Gene Therapy, Protein Therapy, and PTC124

Gene therapy trials for treatment of hemophilia, using recombinant factor VIII or factor IX cloned into adeno-associated vectors, were described by Pierce et al. (2007). These investigators noted that worldwide there are approximately 400,000 patients with hemophilia and that 300,000 likely receive sporadic treatment or no treatment. Treatment with factor VIII or factor IX protein is very costly and frequently induces neutralizing antibodies. One advance in the use of protein for replacement therapy is the development of the combination of coagulant factor VIII or IX with liposomes. This extends the life of the protein.

Pierce et al. (2007) reported that small molecules such as PTC124 that facilitate read through of stop codons have shown promise in the treatment of hemophiliacs with nonsense mutations that constitute stop codons. Approximately 10% of hemophiliacs have such mutations.

Gene Delivery With Viral and Nonviral Vectors

Gene therapy and specifically targeted gene therapy may represent a promising strategy for treatment of cancer. In some cases, the goal may be the introduction of a tumor suppressor gene. In other cases, therapy with antiangiogenesis factor or angiostatic factors may be valuable (Brandwijk et al. 2006). Adeno-associated viruses may be preferred vectors since they are less immunogenic. Nonviral delivery systems used to deliver DNA to tumors include lipid-based systems such as liposomes and cationic lipid complexes, or polymers such as polyethylenimine. Bondi et al. (2007) noted that these vehicles protect DNA from degradation and promote uptake into cellular endosomes.

TISSUE-TARGETED GENE THERAPY: SUBTHALAMIC INJECTION OF ADENOVIRAL VECTOR CONTAINING THE GLUTAMIC ACID DECARBOXYLASE GENE IN PARKINSON DISEASE

Kaplitt et al. (2007) reported results of an open label safety and tolerability trial of unilateral subthalamic injection of adeno-associated viral vector containing the glutamic acid dehydrogenase gene (AAV-GAD). They reported that patients were followed for a period of up to 12 months following therapy. Results indicated significant improvement in motor function and control of tremors on the side of the body contralateral to the subthalamic nuclear injection site.

Parkinson disease is associated with degeneration of neurons in a number of different brain regions. However, the most characteristic defect is loss of dopaminergic neurons of the substantia nigra leading to movement disorders, rigidity, and tremor. Dopamine-related therapies are useful in the early stages of the disease but may actually worsen symptoms later. Kaplitt et al. (2007) used gene therapy with the glutamate dehydrogenase gene that catalyzes synthesis of GABA (gamma-aminobutyric acid) based on prior evidence that GABA infusions can ameliorate disease manifestations in Parkinson disease. They continued dopaminergic treatment following unilateral subthalamic injection of AAV-GAD.

Adeno-associated gene therapy trials are in progress for other neurological diseases, including Batten disease and Canavan disease. Kaplitt et al.

noted that targeted gene therapy in the brain is important because vectors most often do not cross the blood-brain barrier.

DEMONSTRATION OF POTENTIAL FOR TREATMENT OF FRAGILE X MENTAL RETARDATION BASED ON STUDIES IN MOUSE MODELS

The syndrome that has become known as fragile X mental retardation was first described by J. P. Martin and Julia Bell in 1943. The syndrome occurs primarily in males and is associated with mental retardation, autism, and a fairly distinct pattern of facial features: long face, large ears, and high-arched palate. This syndrome was referred to as the Martin Bell syndrome prior to the discovery of the abnormal X chromosome, "a marker X chromosome," in cytogenetic studies by Lubs (1969).

Fragile X syndrome (FXS) is due to transcriptional silencing of the *FMR* gene. Transcriptional silencing is due to triplet repeat expansion in the 5' region of the fragile X mental retardation gene (*FMR1*; Oberle et al. 1991). Symptoms vary depending on the degree of repeat expansion and may include developmental delay, anxiety, attention deficit hyperactivity, stereotypic behaviors, and problems with social interactions.

Development of a Mouse Model and Studies of Pathophysiology in FMRP Deficiency

O'Donnell and Warren (2002) developed a mouse model of this syndrome through knockout of the *FMR1* gene. This model is useful in investigations of the effects of loss of *FMR1*. Studies in neuropathology and neurophysiology indicate that deficits in the development of dendritic spines and abnormalities of synaptic plasticity are key factors in the development of FXS symptoms. The dendritic spine abnormalities in patients with FXS and in the mouse model are similar; there are increased numbers of long immature spines and evidence of impaired signal transmission at glutamatergic synapses.

A number of different studies have demonstrated that the FMR protein (FMRP) binds to mRNA and plays a key role in translation of mRNA that encodes specific proteins. Khandjian et al. (2004) demonstrated that FMRP is present in polyribosomes that are actively translating mRNA. These authors noted that RNA binding proteins play a key role in the transcriptional regulation of gene expression; within the RNA binding proteins there are RNA recognition domains. A number of studies including those of Schaeffer et al. (2001) indicate that FMRP acts as a negative regulator of translation.

Evidence for Impaired Synaptic Function and Potential for Therapy

Huber et al. (2002) reported that the translation of mRNA of metabotropic glutamate receptor (mGluR) RNA and FMRP concentration are correlated. They reported that excess mGluR function and the associated excess in long-term depression of synapses occur in the absence of FMRP and lead to the behavioral phenotype in FXS. They proposed further that antagonists of mGluR be considered as therapeutic agents. Specifically, they proposed that drugs that inhibit group I mGluR and drugs that inhibit long-term depression (LTD) of synaptic function might be useful in the treatment of FXS.

There is evidence that protein synthesis-dependent LTD of synaptic activity occurs not only during development but also later in life. Furthermore, this LTD occurs not only in the hippocampus but also in the cortex and the cerebellum. Studies by Bear et al. (2004) revealed that exaggerated LTD led to increased synaptic loss during critical periods of development.

FMRP and X-Linked P21 Activated Kinase

Studies by Lee et al. (2003) revealed that FMRP binds to the mRNA that encodes P21 activated kinase (*PAK3*), one of the main downstream effectors of RacGTPase. Interestingly, X-linked *PAK3* gene mutations occur in some forms of mental retardation (Allen et al. 1998).

Transgenic mice in which *PAK* activity is abnormal have abnormal dendritic spines, and Hayashi et al. (2007) reported that the spine abnormality is opposite to that seen in *FMR*. There are few spines and a lower proportion of the spines are long and thin. They noted that *Pak1* plays a key role in actin polymerization and dendritic spine morphogenesis. Based on these observations, Hayashi et al. (2007) carried out studies to determine whether mutations leading to *PAK* deficiency could rescue the phenotype in *FMR1* knockout mice. They crossed transgenic mice deficient in *PAK* (dnPAKTG) with *FMR1* knockout mice. In offspring deficient in both *PAK* and *FMR1*, they determined that spine length and density approximated that observed in normal mice and long-term potentiation was indistinguishable from normal. They noted that in the dnPAKTG × FMR1KO mice there was also rescue of behavioral deficits.

Hayashi et al. (2007) then carried out studies to determine whether or not *Pak1* and FMRP interact at synapses. *Pak1* immune precipitates of synaptic membrane proteins were found to contain FMRP. They identified *Pak1* binding sites in FMRP using a series of different FMRP gene deletions.

These studies are of interest in design of therapeutic interventions in FXS since a number of different *PAK* inhibitors have been developed. Nheu et al. (2002) developed two *PAK* inhibitors from naturally occurring ATP antagonists CEP1343 and KTD606. They noted that indolocarbazole compounds may also be useful for treatment of *Pak1*-induced cancers.

Splicing Mutations and Nonsense Mutations: Therapeutic Possibilities

Possibilities for therapeutic interventions to treat disorders due to splicing mutations and nonsense mutations are discussed in Chapter 3.

7

FROM GENE DISCOVERY IN GENETIC SYNDROMES TO IDENTIFICATION OF THERAPEUTIC AGENTS

Elucidation of the molecular defects in a number of rare genetic syndromes characterized by abnormal cell proliferation and differentiation have led to the discovery of genes that were subsequently shown to play a major role in cell signaling and metabolism and also in the development of sporadic tumors not related to the genetic syndromes. Discussed in this chapter are the genes that play a role in tuberous sclerosis, polycystic kidney disease, neurofibromatosis, and Peutz-Jeghers syndrome and translational aspects of gene discovery.

TUBEROUS SCLEROSIS

Tuberous sclerosis is a disorder characterized by abnormal cell proliferation and differentiation, giving rise to hamartomas in a number of different organs including kidney, liver, heart, and brain. The disorder may also lead to seizures and cognitive impairment. Tuberous sclerosis is inherited as an autosomal dominant; however, in at least 50% of cases there is no family history and the disorder is likely due to new mutations. Mutations in either one of two different genes give rise to the clinical manifestations. The *TSC1* gene that maps to chromosome 9q34.1 encodes hamartin; the *TSC2* gene on chromosome 16p13.3 encodes tuberin (Sampson 2003).

Relationship of the Gene Products Tuberin and Hamartin

The proteins encoded by *TSC1* and *TSC2*, hamartin and tuberin, form a physical and functional complex. Hodges et al. (2001) reported that the region encompassed by amino acid 302-430 in hamartin interacts with amino acid 1-418 in tuberin, that is, with the N-terminal region of tuberin. In patients with tuberous sclerosis who have mutations in these regions, the molecular interactions are dramatically affected.

Benvenuto et al. (2000) reported results of experiments that suggested that the *TSC1* gene product hamartin stabilizes tuberin. They also reported that tuberin is highly ubiquinated and is subject to proteosome degradation. Hamartin is less ubiquinated.

Studies in *Drosophila* first demonstrated that phosphorylation of tuberin inhibits its function (Potter et al. 2002). Dan et al. (2002) reported that *PI3K* (phosphatidylinositol-3-kinase) or *AKT* (protein kinase) induce tuberin phosphorylation within the hamartin tuberin complex. They noted that this phosphorylation directly affected the function of the tuberin hamartin complex. The inhibition of tuberin function by phosphorylation likely depends upon the fact that phosphorylation reduces the activity of the Rheb (Ras homolog enriched in brain) GTPase domain of tuberin.

Rheb and Its Key Role in Tuberin-Hamartin Function

Rheb is a highly conserved member of the Ras superfamily of G proteins. These proteins play a role in signaling and growth and are present in a complex either with GDP or GTP. The transition from the "on" (GTP associated) to the "off" form (GDP associated) requires additional proteins, GTPase activating proteins (GAP). Subsequent replacing of GTP requires guanine nucleotide exchange factors (GEFs).

A specific domain of tuberin functions as a GAP for Rheb; it therefore promotes hydrolysis of Rheb GTP to Rheb GDP (Li et al. 2004). In *TSC2*-deficient cells, levels of Rheb GTP are increased. The downstream effect of overexpression of active Rheb is to stimulate phosphorylation of S6K (S6 kinase) and 4EBP1 (eukaryotic translation initiation factor 4E binding protein 1). A protein kinase encoded by the gene *mTOR* (mammalian target of rapamycin) acts as the intermediary molecule between Rheb and its downstream effects.

The intact functional *TSC1 TSC2* complex stimulates hydrolysis of Rheb, decreases activation of *mTOR* and therefore decreases S6 and 4EBP1 phosphorylation. This in turn leads to decreased mRNA translation and decreased protein synthesis.

Manning and Cantley (2003) reported that the *mTOR* and phosphoinositide-3-kinase (*PI3K*) and *AKT* (serine threonine protein kinase) pathways interact. The *PI3K* pathway is activated by mitogenic stimuli while the *mTOR* pathway is activated by mitogenic stimuli when sufficient nutrients are present. The coordinated activity of these two pathways results in activation of mRNA translation and protein synthesis. This in turn results in cell growth.

El-Hashemite et al. (2003) reported definitive evidence for a key role for the *mTOR* pathway in tuberous sclerosis. They analyzed angiomyolipomas in patients with tuberous sclerosis and *TSC2* defects and determined that levels of phosphorylated S6 kinase were increased (Figure 7–1).

mTOR

TOR genes and their products were first identified in yeast by Heitman et al. (1991) in a study of factors that determine cell cycle arrest following use of the immunosuppressant rapamycin.

mTOR is a protein kinase. Inoki et al. (2005) reported that *mTOR* function is enhanced by growth factors and nutrients, especially amino acids and glucose. *mTOR* function is inhibited by rapamycin. There is evidence that *TOR* exists in complexes. *TORC1* is a complex of mTOR with two other proteins, including Raptor and mLST8. *mTORC2* contains the protein rictor *TORC1* is sensitive to inhibition by rapamycin. Rheb is a negative regulator for *mTORC1*.

Rapamycin

Rapamycin was isolated from a species of bacteria, *Streptomyces hygroscopicus*, in soil obtained on Easter Island (Rapa Nui), in a project designed to identify new antifungal agents. It was shown to have immunosuppressant functions (Pritchard 2005). Rapamycin (also known as sirolimus) has an intracellular receptor that binds directly to *mTOR*, specifically to *mTORC1*. Following this binding, the downstream effects of *mTORC1* on S6 kinase and 4EBP1 are suppressed.

Analogs of rapamycin with greater solubility than the parent molecule have been developed and approved for cancer treatment. These include temsirolimus and everosirolimus. Response of *mTOR* to rapamycin differs in degree in different tumors. Sabatini (2006) reported that tumors that benefit from

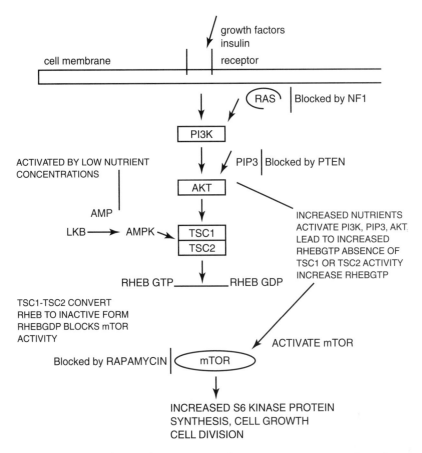

Figure 7-1. Signaling pathways activated in response to nutritional conditions converge on *TSC1, TSC2*, Rheb, and *mTOR*. Absence of *TSC1* or *TSC2* gene products leads indirectly to activation of *mTOR*, and increased cell growth and cell division. Rapamycin blocks the effect of *mTOR* (modified from Inoki et al. 2005).

rapamycin treatment include tumors (hamartomas) characterized by loss of *TSC1* or *TSC2*, and tumors with loss of *PTEN* and activated *AKT* signaling. Tumors with marked angiogenesis due to *VEGF*-driven angiogenesis, abundant *HIF1alpha* expression or *VHL* (Von Hippel Lindau) gene loss and tumors with cyclin overexpression, such as mantle cell lymphoma, may be successfully treated with rapamycin or its analogs.

Inhibitors of the *mTOR* pathway that are not rapamycin related are under development. These drugs are being sought since rapamycin does not completely inhibit tumor growth.

Rapamycin and the Treatment of Hamartomas in Tuberous Sclerosis

Wienecke et al. (2006) and Herry et al. (2007) have reported dramatic effects of *mTOR* inhibitors (rapamycin/sirtulin) on the size of renal angiomyolipomas in tuberous sclerosis patients. The reduction in tumor size led to significant clinical improvement and pain reduction. In the case described by Herry et al., treatment was continued for 2 years and was well tolerated. CT imaging demonstrated dramatic decrease in tumor size. Furthermore, tumors did not recur during a 6-month period following discontinuation of treatment (Figure 7–2).

Franz et al. (2006) reported results of their studies on the treatment of subependymal giant cell astrocytomas in tuberous sclerosis patients. These tumors occur in 5% to 15% of patients with tuberous sclerosis. They often lead to serious neurological complications due to mass effects and obstruction of cerebrospinal fluid flow. Surgical treatment of these tumors is often fraught with serious postoperative complications. Franz et al. treated five tuberous sclerosis patients with brain tumors, four with subependymal giant cell astrocytomas and one with pilocytic astrocytoma. Tumors were visualized with magnetic resonance imaging before and during therapy. Franz et al. reported that in all cases, rapamycin treatment led to tumor regression.

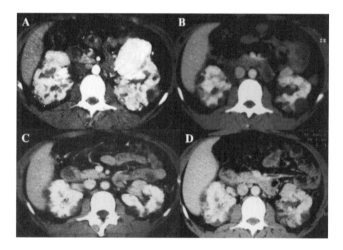

Figure 7–2.   Sirolimus (a modified form of rapamycin) treatment has a dramatic effect on angiomyolipomas in tuberous sclerosis. Computed tomography of the kidneys disclosing angiomyolipomas (A) before treatment, (B) after 1 year of treatment, (C) after 2 years of treatment, and (D) 6 months after the end of treatment with sirolimus (Herry et al. 2007; used with permission).

Taille et al. (2007) reported successful treatment of lymphangioleio-myomatosis with rapamycin. This condition is one of the manifestations of *TSC1* or *TSC2* deficiency.

Other Potential Therapeutic Agents for *TSC* Tumors

Hsu et al. (2007) identified *TCTP* (translationally controlled tumor protein), a molecule that acts as a direct regulator of Rheb. In their studies in *Drosophila*, Hsu et al. determined that *TCTP* interacts with nucleotide free Rheb and stimulates guanine nucleotide exchange and Rheb activation. They propose that *TCTP* is a potential therapeutic target for tuberous sclerosis. Inactivation of *TCTP* may impair Rheb function. There is evidence from other studies that reduction in *TCTP* can suppress malignancy and bring about tumor reversion (Tuynder et al. 2002, 2004).

Additional Functions for Hamartin and Tuberin

There is evidence that hamartin plays a role in apoptosis through its interaction with other proteins (Yasui et al. 2007). Rosner et al. (2007) reported that *AKT* is one of several kinases that phosphorylate tuberin. These include 5' AMP-activated protein kinase, Map kinase, and ERK 1 and 2. As a result of phosphorylation, tuberin loses its ability to act as a GTPase for Rheb.

Rosner et al. noted that an additional activity of tuberin, the capacity to regulate p27, is separate from its *mTOR*/S6 function. Dan et al. first reported this function of tuberin-hamartin in 2002. *AKT* specifically phosphorylates amino acids S939 and T1462 in tuberin. Rosner et al. (2007) reported that the location of tuberin in the cell is affected by the degree to which it is phosphorylated by *AKT*. In the presence of high *AKT* levels and in cycling cells, it is located in the cytoplasm predominantly. In G0 resting cells, *AKT* concentration is low and tuberin occurs primarily in the nucleus. Following phosphorylation of nuclear tuberin, it moves to the cytoplasm.

POLYCYSTIN, TUBERIN, AND THE *MTOR* PATHWAY IN POLYCYSTIC KIDNEY DISEASE

Autosomal dominant polycystic kidney disease type 1 (ADPKD1) is due to mutations in the gene encoding a membrane spanning protein polycystin, PC1. This protein spans the renal epithelial membrane multiple times and has a cytoplasmic tail of 226 residues. The function of PC1 is poorly understood.

Shillingford et al. (2006) reported that the cytoplasmic tail of polycystin interacts with tuberin. They demonstrated further that in renal epithelial cells from ADPKD1 patients the *mTOR* pathway is inappropriately activated. They reported that in ADPKD1 kidney transplant patients where one kidney was removed and patients were treated with rapamycin as an immunosuppressant, there was significant reduction in size of the remaining polycystic kidney.

Shillingford et al. demonstrated inappropriate activity of the *mTOR* pathway in ADPKD1 kidney sections through antibody against active forms of *mTOR* and S6 kinase.

LKB1 TUMOR SUPPRESSOR KINASE AND PEUTZ-JEGHERS SYNDROME

Peutz-Jeghers syndrome is an autosomal dominant disorder characterized by the development of hamartomas, primarily polyps of the gastrointestinal tract, and unusual pigmentation involving oral mucosa, face, palmar surfaces, and genitalia. The hamartomatous polyps may lead to symptoms in the second or third decade of life, including bleeding, occlusion, and intussusceptions. Patients with this syndrome have an 18-fold increase above population risk of developing gastrointestinal cancers. They are also at increased risk for other cancers, including breast, pancreas, and ovarian cancers.

The Peutz-Jeghers syndrome was mapped to chromosome 19p13.3 and shown to be due to mutations in a serine threonine kinase gene *STK1*, also known as *LKB1* (Jenne et al. 1998). This kinase is unusual in that it acts as a tumor suppressor. However, studies carried out on human polyps indicate that biallelic inactivation of *LKB1* is not necessary for hamartoma formation (Katajisto et al. 2007). The *LKB1* gene product interacts with at least 14 different kinases and plays a role in a number of signaling pathways, including those involving AMPK (5′ adenosine monophosphate kinase) and the PAR (polarity regulation pathway). Studies by Sanchez-Cespedes et al. (2002) revealed that *LKB1* mutations occur in 30% of sporadic lung adenocarcinomas.

Katajisto et al. reviewed the relationship of *LKB1* to the function of AMPK. This enzyme is activated by an increase in intracellular adenosine monophosphate (AMP) that occurs concurrently with a decrease in adenosine triphosphate (ATP). Under these conditions, a conformational change occurs in 5′ AMPK due to the binding of AMP to its gamma subunit. Subsequently, *LKB1* phosphorylates the alpha subunit. AMPK in turn phosphorylates a number of metabolic enzymes, leading to shutdown of glycogen synthesis and lipid synthesis, as the cell shifts to an energy-releasing mode. This AMPK pathway interacts with the *TSC1-TSC2-TOR* signaling pathway. Wild-type *LKB1*, through AMPK and its effects on *TSC1* and *TSC2*, inhibits the *mTOR* pathway. There is evidence that *mTOR* shows increased activity when *LKB1* function is reduced through mutations, as in Peutz-Jeghers syndrome. Given the activation of *mTOR* in *LKB1*-deficient tumors, inhibitors of the *mTOR* pathway such as rapamycin may play a role in the treatment of Peutz-Jeghers syndrome (Figure 7–1).

There is also evidence that in the hamartomatous polyps in Peutz-Jeghers syndrome, *COX2*, which catalyzes prostaglandin synthesis, is overexpressed and plays a role in polyp formation. Clinical trials with *COX2* inhibitors have led to a reduction in polyp size. Other drugs in clinical trial for this syndrome include drugs that directly impact AMPK activity. These drugs, metaformin and phenformin, are used to treat type 2 diabetes. However, these drugs require that at least one allele of *LKB1* is present.

NEUROFIBROMATOSIS: ONGOING RESEARCH AND ITS THERAPEUTIC RELEVANCE

Neurofibromatosis is characterized by the development of tumors, primarily neurofibromas that arise in the skin and in the peripheral and central nervous system. Pigment changes, café au lait spots, and axillary freckling are common in this disorder. Eye changes (Lisch nodules) and bony changes (erosions) are often present. Vascular changes may occur, leading to strokes and cerebral ischemia. The population frequency of neurofibromatosis (*NF1*) is 1 in 3000. It is inherited as an autosomal dominant condition; however, 30% to 50% of cases arise as new mutations (Radtke et al. 2007).

The *NF1* gene maps to chromosome 17q11.2 and encodes neurofibromin (Gutmann and Collins 1993). Disease-causing mutations occur throughout the gene. They may be single nucleotide mutations or deletions. Identification of the causative mutation in a specific *NF1* patient is hampered by the large size of the gene, the fact that mutations may arise throughout the gene, and the presence of many *NF1* pseudogenes elsewhere in the genome. It is interesting to note that children with features of *NF1* and with hematological disorders have been found to have mutations in the mismatch repair gene *MLH1* (Wang et al. 2003) and *MSH6* (Hegde et al. 2005).

NF1 Tumors and Molecular-Based Treatment

In individuals with germline *NF1* mutations, tumors that develop often show inactivation of the wild-type allele. Neurofibromas may undergo malignant transformation to yield peripheral nerve sheath tumors, which often arise in plexiform neurofibromas. These are tumors that arise in nerve sheaths that surround peripheral nerves, large nerve trunks, or spinal nerve roots. The incidence of astrocytomas and gliomas is higher in patients with neurofibromatosis than in the general population. Malignant transformation in these tumors is associated with marked chromosome instability both in chromosome copy and instability at microsatellite loci.

Dasgupta and Gutmann (2005) reviewed the molecular genetics of the neurofibromin gene and signal transduction changes in *NF1*. The neurofibromin gene contains a small domain that is homologous to RAS GTPase (RAS, regulatory protein; GTP, hydrolyzing proteins). These enzymes catalyze conversion of the active form of RAS, RASGTP, to the inactive form, RASGDP. In *NF1* tumors there are high levels of active RAS due to decreased RASGAP function. Initial therapies of *NF1*-associated tumors involved the use of anti-RAS agents. RAS is active in membranes and requires isoprenylation for its activity. Early therapies involved the use of farnesyl transferase inhibitors that impact isoprenylation. Although these substances worked well in *NF1*-deficient cultured cells, they were not useful for inhibition of tumor growth in patients.

Further experimentation on neurofibromin revealed its role in other signal transduction pathways. Dasgupta and Gutmann reported that *PI3K*-dependent activation of *mTOR* signaling is enhanced in *NF1* tumors. Furthermore, there is evidence that KRAS rather than HRAS may be the key target of neurofibromin signaling. In addition, neurofibromin plays a role in the protein kinase C pathway (PKC). PKC phosphorylates *NF1* and increases its interaction with the actin cytoskeleton.

Dasgupta et al. also reported that proteins involved in ribosome biogenesis were upregulated in *NF1* tumor cells. Loss of *NF1* from these cells led to S6 kinase upregulation and increased protein synthesis. They noted that treatment with rapamycin inhibited S6 kinase–related hyperproliferation. Several rapamycin analogues are now being used in clinical trials. They include CCL779.

Neurofibromas May Occur as One Feature of an Unusual Syndrome

Neurofibromatosis manifestations may occur as features in a contiguous gene syndrome due to a large deletion on chromosome 17q11.2 (Riva et al. 1996).

Recent clinical studies revealed that specific mutations in the *NF1* gene may lead to features of Noonan syndrome, with evidence of congenital heart disease. Huffmeier et al. (2006) postulated that these gene mutations and mutations in genes, at other locations that cause Noonan syndrome, interrupt Ras pathway function (discussed further in Chapter 11).

8

IMPACT OF THE ENVIRONMENT ON THE GENOME

Humans are embedded in Nature. The biologic science of recent years has been making this a more urgent fact of life. The new hard problem will be to cope with the dawning intensifying realization of just how interlocked we are.

—Lewis Thomas, *The Lives of a Cell*, 1978

Changes in DNA chemical composition, integrity, or structure may be induced through endogenous or exogenous factors, the latter of which may be chemical or physical factors. Genomic instability may occur as a direct or indirect result of environmental factors. The interactions between the environment and germinal tissues (ovary, eggs, testes, and sperm) and interaction between the environment and the developing organism during critical periods of vulnerability are of particular concern. Insight into agents that damage DNA and into mechanisms of DNA repair have been in part derived through studies on hereditary disorders characterized by a tendency to develop tumors.

DNA DAMAGE AND REPAIR: INSIGHTS
FROM HEREDITARY DISORDERS

DNA-damaging factors include endogenous factors, such as reactive oxygen species and particular cellular metabolites, and exogenous factors including ultraviolet light, ionizing radiation, and toxins. DNA adducts are chemicals that complex to DNA bases.

Proteins involved in DNA repair offset the effects of DNA-damaging agents. Different types of DNA repair include base excision repair, nucleotide excision repair, and mismatch repair (Friedberg 2003).

Base excision repair (BER) occurs when nucleotides are altered through endogenous oxidation or alkylation. An important source of mutation is de-amination of 5-methyl-cytosine that leads to the formation of thymine. Damaged bases are cleaved from the sugar phosphate backbone of DNA by gly-cosylases. Endonucleases incise the damaged DNA strand and the DNA strand is repaired by DNA synthesis and ligation. Base excision repair may involve short patch repair or long patch repair. At least eight different glycosylases play a role in BER. Biallelic mutations in the glycosylase MYH, also known as MUTYH, are associated with defects in BER and clinically with a syndrome characterized by development of colorectal adenomas and carcinomas. This glycosylase is involved in the excision of oxidized guanosine (Lu et al. 2001).

Nucleotide excision repair (NER) is a different process designed to re-move pyrimidine adducts. The NER process involves removal of a number of nucleotides. This process requires unwinding of the DNA duplex and incision of the flanking DNA, followed by DNA synthesis. Nucleotide excision repair defects occur in the inborn error syndromes xeroderma pigmentosum, Cock-ayne syndrome, and trichothiodystrophy. Xeroderma pigmentosum is char-acterized by extreme sensitivity to sunlight. It is due to deficiency of any one of seven different proteins involved in NER. In Cockayne syndrome, NER defects lead to growth retardation and neurological deficits.

Mismatch repair (*MMR*) removes incorrect nucleotides that were inserted during replication. Factors recruited to the site degrade the mismatch and this is followed by DNA synthesis. In hereditary nonpolyposis colorectal cancer (HNPCC), germline mutations occur in *MMR* genes. HNPCC is an autosomal dominant condition. Adenomas do not occur. However, patients develop colon tumors characterized by novel alleles at microsatellite markers. The micro-satellite instability, characterized by variable lengths of microsatellite repeat polymorphisms, arises as a result of mutations in a mismatch repair gene, primarily *MSH1*, *MSH2*, or *MSH6* (Fishel 1999). Mismatch repair is discussed further on pages 122 and 123.

Double-stranded DNA breaks may be repaired through homologous re-combination. This process requires that a sister chromatid be available and so is restricted to the S or G2 phase of the cell cycle. The sister chromatid serves as a template for DNA synthesis repair and the process is considered to be error free. End joining may also repair double-stranded DNA breaks. This process results in loss of nucleotides or in translocations. A large number of different proteins bind to double-stranded breaks. A number of these proteins play a role in controlling the cell cycle, so that cell division does not occur before the defect is repaired.

Insight into mechanisms of DNA damage and repair has arisen through studies of rare genetic syndromes characterized by abnormal sensitivity to DNA-damaging agents or abnormal DNA repair processes.

Targeting Proteins to Sites of DNA Damage: Fanconi Anemia

Fanconi anemia is a recessively inherited disorder. Mutations in any one of at least eight different genes lead to this disorder. Patients with Fanconi anemia develop progressive bone marrow failure. In addition, Fanconi anemia patients may have congenital malformations. They may develop hematological malignancies and tumors (Mathew 2006). Characteristic features of the syndrome are hypersensitivity to DNA-oxidizing agents and DNA cross-linking agents. Six of the eight genes involved in Fanconi anemia encode proteins that form a multisubunit nuclear core complex. Proteins within this complex function in a pathway that determines cellular response to DNA damage (Risinger and Groden 2004; Figure 8–1). This cellular response results in targeting of specific Fanconi proteins to sites of DNA damage. Interstrand cross-link repair requires both nucleotide excision and homologous recombination (see also Chapter 11).

RecQ-like DNA Helicases: Bloom, Rothmund-Thomson, and Werner Syndromes

RecQ helicases play a role in DNA replication, repair, and recombination and are therefore important in maintenance of genomic stability. At least five different genes in the human genome encode RecQ-like helicases. Mutations in three of these genes occur in autosomal recessive chromosome breakage syndromes, including Bloom syndrome, Rothmund-Thomson syndrome, and Werner syndrome. Each of these syndromes is characterized by increased susceptibility to malignancy. Short stature and extreme sensitivity to sunlight are features of Bloom syndrome. The cellular phenotype is characterized by a

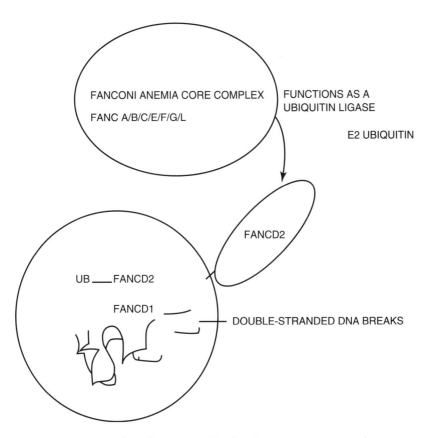

Figure 8–1. Genes that when mutated lead to Fanconi anemia encode a series of proteins that form a core complex. This core acts as a ubiquitin ligase and targets the *FAND2* gene product to the nucleus. *FAND2* and *FAND1* gene products interact and are targeted to double-stranded DNA breaks and play a role in repair of these breaks (based on Risinger and Groden 2004).

high frequency of exchange between sister chromatids. This exchange results in homologous recombination. Bloom syndrome protein has a punctate distribution in nuclei; it is also found in PML bodies (these bodies are prominent in promyelocytic leukemia). Bloom syndrome protein associates with sites of DNA damage and with the stalled replication forks that occur when DNA is damaged.

Risinger and Groden (2004) noted that the main function of RecQ helicases may be to restart replication at stalled forks. RecQ helicase deficiency leads to aberrant DNA replication and a high frequency of recombination. These investigators noted that further studies are required to determine whether subtle sequence variants in DNA repair genes play a role in the pathogenesis of more common cancers. Understanding the role of such defects in the origin of a specific tumor may guide selection of therapeutic agents. *WRN*, the gene that is defective in Werner syndrome, encodes a form of RecQ helicase that serves to prevent telomere loss. Early telomere loss occurs in Werner syndrome and leads to premature aging (Crabbe et al. 2007).

Checkpoint Genes and Cell Cycle: Ataxia Telangiectasia and Nijmegen Breakage Syndrome

Double-stranded DNA breaks are common in ataxia telangiectasia (ATM) and in Nijmegen breakage syndrome (NBS). Characterization of the genes responsible for these disorders has provided insight into mechanisms of cell cycle control that come into play when DNA is damaged. Double-stranded chromosome breakage is followed by phosphorylation of histone H2X at that site. *ATM* and *ATR* (ATM related) serine threonine kinases carry out phosphorylation. Histones therefore play an important role in DNA repair. Following phosphorylation, DNA repair proteins are recruited to the breakpoint.

Cell cycle checkpoints are mechanisms that come into play to slow down or halt the cell cycle to allow repair of DNA damage resulting from exposure to endogenous or exogenous genotoxic agents. DNA damage that leads to extensive strand breaks triggers nuclear mechanisms that result in recruitment of proteins to the site of the breaks. The downstream effect of this recruitment is arrest of the cell cycle (Houtgraaf et al. 2006).

Sensors of DNA damage are proteins encoded by *RAD* (radiation related) and *NBS1* genes. These interact with proteins encoded by the *ATM* and *ATR* genes. The *CHK1, CHK2, p53*, and *p21* genes encode other important checkpoint proteins. Arrest in the G1/S phases of the cell cycle serves to prevent replication of damaged DNA. Arrest in the G2/M phases of the cell cycle prevents segregation of damaged chromosomes into new cells. Cell cycle checkpoint mechanisms may also trigger cell death (apoptosis). *CHK1, CHK2, ATM*, and *ATR* are serine threonine kinases that phosphorylate targets and then lead to cell cycle arrest.

MICROSATELLITE INSTABILITY

Forms of colon cancer that are associated with microsatellite instability may also show defects in *CLK1*, a gene encoding a kinase that is involved in checkpoint control. Microsatellite instability may result from a number of different DNA *MMR* gene defects, such as mutations in *MSH1, MSH2, MSH6*, and *PMS2*, and from defects in genes involved in double-stranded DNA repair, such as RAD 50.

Genetic screening for *MMR* defects in individuals with a family history of colon cancer is based on the premise that disease risk and mortality may be reduced by the application of frequent colonoscopy (Barnetson et al. 2006). The lifetime risk ratio for colorectal cancer in males with *MMR* defects is 83 and for females it is 48. Defects in the *MSH1, MSH2*, and *MSH6* genes are predominant in hereditary nonpolyposis coli. There is evidence that other genes may be involved.

MMR gene screening is recommended for individuals in families that meet the Amsterdam II criteria: family history positive for colon cancer, three affected individuals over two generations, one individual younger than 50 at the time of development of cancer, and a first-degree relative positive for colon, endometrial, small bowel, or upper urethral tract cancer.

Colon cancer patients younger than 50 years of age who do not meet the Amsterdam criteria frequently have *MMR* gene defects, and it is important that these individuals be screened (Mead et al. 2007). These investigators reported that in 59% of cases where microsatellite instability was found in tumors, the family did not meet Amsterdam criteria. Furthermore, they noted that the microsatellite instability test should include not only analysis of dinucleotide repeats but also analysis of mononucleotide repeats. This form of microsatellite instability occurs particularly in association with *MSH6* mutations

Other investigators have emphasized that mismatch repair gene mutations lead to instability of microsatellite polymorphisms and that tumors should be screened for these, (Ionov et al. 1993). Immunohistochemical screening of MMR proteins in tumors is sometimes carried out.

CHROMOSOME INSTABILITY: TELOMERE SHORTENING AND ABNORMAL SEGREGATION

Alterations in dosage levels of proteins that play a role in DNA synthesis and repair may precipitate chromosome instability (Bayani et al. 2007). It is, however, not clear whether copy number instability drives malignancy or whether it arises in consequence of the malignant process.

Fridlyand et al. (2006) reported results of studies of genome dosage in breast cancer. They observed that in tumors in which telomeres were shortened, chromosome copy number changes were most common. They proposed that telomere aberrations lead to aberrant end joining and to breakage fusion bridge processors. They reported that *BRCA1* mutations are associated with genome instability and poor survival.

Familial adenomatous polyposis is an autosomal dominant disorder due to germline mutations in the *APC* (adenomatous polyposis coli) gene. This disorder is characterized by the presence of hundreds of colorectal adenomas. A number of these may undergo malignant transformation in consequence of somatic mutation of the wild-type *APC* allele. *APC* mutations also predispose to desmoid tumors and to hypertrophy of the retinal pigment epithelium. The *APC* gene plays a role in chromosome segregation. Defects in chromosome segregation lead to abnormal chromosome number and genomic imbalance.

INTERINDIVIDUAL VARIATION IN DNA REPAIR

Maintenance of genomic integrity is a key element in cancer prevention. Wu et al. (2007) noted that there is considerable interindividual variation in capacity to repair DNA damage. One method to assess mutagen sensitivity is based on quantitation of chromatid breaks in blood lymphocyte cultures in response to mutagen treatment at a defined dose. Validation of this test as an indicator of ability to repair DNA damage was derived from studies of inherited defects of DNA repair. Mutagens used in these tests include bleomycin, tobacco-related hydrocarbons, and ultraviolet light.

Wu et al. (2007) reviewed results of prospective studies. They reported evidence that individuals whose lymphocytes manifested high sensitivity to bleomycin had a greater risk that a precancerous lesion would evolve into cancer. For example, patients with Barrett's lesions of the esophagus and evidence of increased lymphocyte sensitivity were more likely to develop esophageal cancer than those with normal degrees of mutagen sensitivity. The risk for development of esophageal carcinoma was even higher if the Barrett's esophagitis lesion manifested loss of heterozygosity of the *p53* gene. These investigators also reported evidence that patients with demonstrated increased sensitivity to mutagen-induced chromatid breaks have a greater chance of developing second primary tumors following treatment of a first cancer.

Wu et al. reported that there is increasing evidence that mutagen sensitivity, as defined in the peripheral blood lymphocyte test, is an inherited trait. Among first-degree relatives of individuals with increased mutagen sensitivity, 62% showed similar increased sensitivity. Increased mutagen sensitivity occurred in only 7% of first-degree relatives of controls with normal sensitivity. Twin studies on monozygotic and dizygotic twins demonstrated that there is a significant genetic contribution to mutagen sensitivity. The degree of heritability differs depending on the mutagen used. It is highest, 68%, in the case of gamma radiation.

One difficulty noted by Wu et al. is that the test as described above requires observer skill. Furthermore, there may be differences in sensitivities between peripheral blood leukocytes and other organs. They propose further that genetic variation is complex as genes may act together to influence mutagen sensitivity. For example, mutations in DNA repair genes and in cell cycle checkpoint genes may influence outcome.

TOXICOGENOMICS

> Along with the possibility of extinction of mankind by nuclear war, the central problem of our age has therefore become the contamination of man's total environment with such substances of incredible potential for harm—substances that accumulate in the tissues of plants and animals and even penetrate the germ cells to shatter or alter the very material of heredity upon which the shape of the future depends. (Carson 1962, p. 8)

Thousands of chemicals of diverse structure and properties are introduced into the environment. The significance of these to health depends on the dosage of each, the timing of exposure, and also genetic diversity. Genetic diversity may exist between humans and animals used to test effects (Waters et al. 2003). Toxicogenomics is a new subdiscipline of toxicology defined by Aardema and MacGregor (2002) as "the study of the relationship between the structure and activity of the genome and the adverse biological effects of exogenous agents" (p. 13). A National Center for Toxicogenomics has been established. The main goal of the center is to develop a reference knowledge base on toxicity and potential toxicity of introduced drugs and chemicals based on gene expression analysis, proteomics, and metabolic profiling.

Aardema and MacGregor (2002) predicted that regulatory and industrial toxicology practices are likely to undergo dramatic changes in consequence of the development of new technologies that examine molecular responses to introduced chemicals. New methodologies will facilitate evaluation and identification of potential toxicants. Furthermore, they suggest that new families of biomarkers will be identified.

Gene expression will be assessed to determine alteration related to toxicant exposure. Fingerprints of exposure may be identified, that is, a specific pattern of gene expression in response to chemicals with specific structural characteristics, or a characteristic pattern of metabolite alterations. Comparison of a fingerprint pattern obtained with an unknown toxicant to patterns observed with a known toxicant will provide information on the toxicity of the unknown compound and ability to predict responses to the compound.

Significant advances in proteomics and metabolomics include multidimensional liquid chromatography, mass spectroscopy, and improved identification of protein sequences using matrix or surface-enhanced laser description ionization techniques (MALDI and SELDI). Specific toxicants exert their effects by impacting protein folding and export. Protein structural changes may also result from oxidation. Changes in proteins through reaction with toxicants may now be more readily analyzed (Aardema and MacGregor 2002).

9

MOLECULAR DEFECTS IN CANCER: APPLICATION TO THERAPY

It is important to consider how understanding of the underlying molecular defects in cancer may be utilized to develop therapies. Many conventional chemotherapy agents target the propensity of cancer cells for unrestricted proliferation. However, treatments with agents that damage DNA or interfere with DNA synthesis have toxic side effects. The margin of difference in impact on cancer and normal cells is small (Adler and Gough 2007). Newer approaches to therapy are often based on exploiting the molecular differences between cancer cells and normal cells. A number of new approaches to therapy that are based on much older observations on differences between cancer cells and normal cells seek to target differences in metabolism, for example, the fact that cancer cells tend to produce adenosine triphosphate (ATP) and energy through glycolysis rather than through oxidative phosphorylation. Pharmacological interventions that reverse this trend in cancer cells stimulate apoptosis of those cells.

The microenvironment in which cancer cells grow is hypoxic relative to that of normal cells. This leads to induction of hypoxia-inducible transcription factors. These transcription factors induce growth factors including EGF (epidermal growth factor) and VEGF (vascular endothelial growth factor). VEGF stimulates angiogenesis. Downregulation of p53 in tumors deprives them of antiangiogenesis factors.

Ideal cancer therapy would be designed to target a specific difference that exists between cancer cells and normal cells so that normal cells are not destroyed. Brody (2005) considered therapies that approach this ideal and included therapies for breast cancer tumors where amplified forms of the receptor ERBB2 (sometimes referred to as HER2, human epidermal growth factor receptor 2) are targeted by antibody, Herceptin, and Gleevec that targets cancer-specific fusion protein, an abnormal tyrosine kinase that occurs in chronic myelogenous leukemia.

Mutations in kinase or in the kinase domains of specific receptor genes constitute important drug targets. EGFR (epidermal growth factor receptor) is a transmembrane receptor tyrosine kinase and is a member of the HER family of receptors. Frequently EGFR is overexpressed by tumor cells or tumor cells develop gain of function mutations. Drugs developed to target EGFR include drugs that inhibit tyrosine kinase activity or drugs that interfere with extracellular ligand binding (Dutta and Maity 2007). Epidermal growth factor binds to the EGFR cell surface receptor that is normally expressed at low concentrations. The N terminal ectodomain binds EGF. There is a 23-amino acid transmembrane domain. The C terminal region encodes a 542 cytoplasmic domain and within this domain there is a region of homology to tyrosine kinase. Binding of ligand to the extracellular ectodomain activates the cytoplasmic tyrosine kinase region. This in turn leads to phosphorylation of tyrosine residues of specific cytoplasmic proteins.

Among lung cancers, 10% have gain of function mutations in the tyrosine kinase domain of the EGFR gene. These lung adenocarcinomas may respond to tyrosine kinase inhibitors geftinimib (Iressa). Pao et al. (2005) reported that despite an initial good response to treatment, patients often acquire resistance to EGFR inhibitors. They determined that in these patients a secondary EGFR mutation arose in exon 20. This was often a T790M mutation.

A novel treatment approach in breast cancer targets a specific weakness in breast cancer cells with *BRCA1* or *BRCA2* mutations. Among their numerous functions, *BRCA1* and *BRCA2* play a key role in the repair of double-stranded DNA breaks by homologous recombination. Inherited mutations of *BRCA1* or *BRCA2* decrease their function. Subsequent mutations or deletions render cells unable to produce proteins that effect DNA repair.

Farmer et al. (2005) investigated the sensitivity of *BRCA*-negative cell lines to inhibitors of the enzyme PARP1 (phosphoadenosine diphosphate ribose polymerase). This enzyme plays a role in the repair of single-stranded DNA breaks. Defects in this pathway result in degeneration of single-strand breaks to double-strand breaks. In cells or in mice deficient in PARP1, the resulting double-stranded DNA breaks are repaired through activity of *BRCA1* or *BRCA2*.

Mice and cell lines with *BRCA1* or *BRCA2* mutations are highly sensitive to loss of PARP1 activity. The combination of severe reduction in *BRCA* and reduction in PARP1 renders cells unable to correct single- or double-stranded DNA breaks. Cell lines accumulate DNA breaks. This leads to chromosome instability, cell cycle arrest, and subsequent apoptosis. These experiments indicate that PARP inhibitors are likely to be useful in treatment of breast cancer due to mutations in *BRCA1* or *BRCA2*. Rubinstein (2007) reported that the potentially high efficacy and low toxicity of PARP inhibitors presents an opportunity for targeted cancer therapeutics for *BRCA1* and *BRCA2* germline mutation carriers.

It is clear that many different genes are mutated or altered in their expression in cancer. It is, however, becoming evident that some mutations, "the drivers," are more important than others, "the passengers," in determining outcome and that therapy may be effective even if it is targeted only to a subset of the mutations (Greenman et al. 2007).

Mutations of p53 are present in approximately 50% of cancers in a variety of different tissues (Vogelstein et al. 2000). In other tumors where p53 is not mutated, its expression is affected indirectly by epigenetic mechanisms (Weisz, Oren, et al. 2007). Therapies to target p53 changes are being actively sought. This is discussed further on pages 139–40.

TUMOR METASTASES

A key question that arises with respect to tumor metastases is whether or not there are patterns of gene expression within tumors that determine their growth in other tissues. Gupta et al. (2007) reported on expression of four genes that contribute to vascular remodeling in primary tumors. These genes also impact entry of tumor cells into the circulation and their subsequent exit into lung tissue. These investigators carried out studies on a cell line derived from the pleural effusion of a patient with widespread metastatic breast cancer. Cultures were analyzed and tumor cells were also transplanted into mice. The tumor cells overexpressed EGFR, Epiregulin, an EGFR ligand, *COX2*, and matrix remodeling metalloproteinases *MMP1* and *MMP2*. The products of these genes are downstream effectors of VEGF in vascular cells.

To downregulate expression of these genes, Gupta et al. used interference RNA. They determined that silencing of the genes had little effect on initial growth of tumor cells transplanted to mammary fat pads in mice. However, decreased expression of all four of the genes led to defects in the vascular morphology of the tumors, particularly in the extent of branching of the vessels and in the size of the vessel lumen. Dye studies revealed decreased vessel permeability. These vascular changes were associated with increased apoptosis of tumor cells. They noted marked differences between control tumor cells and tumor cells in which expression of the genes was knocked down, in the potential for extravasation from capillaries into the lung parenchyma.

Pharmacological targeting of genes with a combination of drugs using the anti-EGFR antibody cetuximab (Erbitux), *COX2* inhibitor celecoxib (Celebrex), and metalloproteinase inhibitor GM 6001 reduced growth of the primary tumor and impaired migration of tumor cells into vessels. In addition, this treatment modified the form of the pulmonary micrometastases. Administration of drugs singly inhibited tumor growth to a minimal extent. Gupta et al. carried out similar studies on freshly isolated breast cancer cells from pleural effusions of metastatic breast cancer cells and obtained similar results.

These studies indicate how biological activities of tumor cells that determine entry into the vasculature and tissue invasion may be specifically targeted.

INFLAMMATORY CHANGES AND CANCER

Balkwill and Mantovani (2001) reviewed the relationship between inflammation and cancer. They noted that in 1863 Rudolf Virchow reflected on the origin of cancer at the site of chronic inflammation. A number of chronic inflammatory conditions are associated with a high risk of cancer. Infections and the cancers they are associated with include: schistosomiasis and bladder cancer; *Helicobacter pylori*, chronic gastritis, and stomach cancer or MALT lymphoma; *Papillomavirus* and cervical cancer; herpes virus type 8 (Kaposi type) and Kaposi sarcoma; Epstein-Barr virus (EBV) and Burkitt lymphoma; pelvic inflammatory disease and serous tumors of the ovary; chronic prostatitis and prostate cancer; chronic esophagitis and Barrett's metaplasia; tobacco smoke–induced chronic bronchitis and lung cancer; inflammatory bowel disease and colorectal cancer.

Pathogenesis of Cancer in Relation to Chronic Inflammation

Leukocytes recruited to the site of infection produce cytokines and growth factors. Among the most important in the pathogenesis of tumors are tumor necrosis factor (TNF), VEGF, and transforming growth factor (TGF). In addition, leukocytes recruited to inflammatory sites produce reactive oxygen species (ROS) and nitric oxide (NO). The latter is generated through cytokine induction of nitric oxide synthase. Nitric oxide and ROS can directly damage DNA and may also repress production and function of DNA repair enzymes (Balkwill and Mantovani 2001).

Helicobacter pylori

Epidemiologic studies demonstrated that *H. pylori* infection is associated with chronic gastritis, gastroduodenal disease, gastric adenoma-carcinoma, and a specific form of lymphoma of the gastric lymphoid tissue (MALT). There are a number of different factors that contribute to the generation of gastric cancer following infection with this organism. Suzuki et al. (2007) considered that the most potent factors that lead to gastric mucosa injury are oxygen-derived free radicals released from activated neutrophils. They report that the chain of events begins with activation of nicotinamide adenine dinucleotide phosphate (NADPH) oxidase in the neutrophil membrane and the subsequent transfer of electrons to molecular oxygen, leading to the formation of superoxide radicals with free electrons. These may then be transferred to hydrogen peroxide or give rise to hydroxyl radicals. Suzuki et al. (2007) noted that activity of neutrophil myeloperoxidase in the presence of hydrogen peroxide and chloride leads to the production of the hypochlorous anion OCl. This may then react with ammonia derived from urea produced by *H. pylori* to produce a monochloramine oxidant NH_2Cl. Reactive oxygen species and oxidants lead to DNA lesions.

Suzuki et al. noted that patients with *H. pylori* infections and high levels of 8-OH-2-deoxyguanosine have a greater degree of gastric atrophy and there are repeated cycles of atrophy and metaplasia. These investigators reported that *H. pylori* infection increases the risk of gastric cancer 10-fold.

Activation-induced cytidine deaminase and mutagenesis in H. pylori infections

Matsumoto et al. (2007) reported evidence for another important mechanism of pathogenesis of gastric cancer due to *H. pylori* infection. They noted that *H. pylori* is a class I carcinogen for human gastric cancer and is classified into pathogenic subtypes cag PAI positive or cag PAI negative. The cag PAI positive forms contain a 41 kb fragment on which 31 genes are located. The cag PAI positive strains produce a much more severe infection and lead more frequently to gastric cancer. These investigators reported that infection with cag PAI positive *H. pylori* induced aberrant expression of the enzyme activation-induced cytidine deaminase (AID). This enzyme functions as an editor of nucleic acid sequences. Normally it is involved in immunoglobulin class switch recombination and mutation, which play a role in the generation of antibody diversity.

Matsumoto et al. proposed that this enzyme plays a key role in tumorigenesis. They investigated the role of overexpression of AID in inducing somatic mutation in gastric cancer associated with *H. pylori* infection. They specifically investigated mutations in p53 in *H. pylori*–associated tumors. They determined that *H. pylori* infection leads to increased expression of the transcription factor NF kappa B. This in turn leads to increased AID expression. They then examined the effects of AID overexpression in the human AGS gastric epithelial cell line and analyzed nucleotide alteration in the p53 gene. They determined that AID overexpression resulted in significant nucleotide alteration in p53. Interestingly, these alterations were clustered in specific exons 2 through 4, 7, and 8. The nucleotide alterations had potential functional significance. Matsumoto et al. therefore demonstrated directly that increased infection with AID leads to p53 mutations (Figure 9–1).

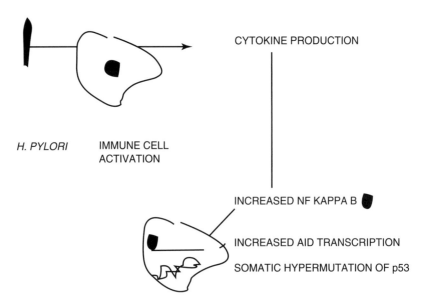

Figure 9–1. *Helicobacter pylori* infection in the gastric mucosa causes an inflammatory reaction with increased production of NF kappa B. This enters the nucleus and increases transcription of the activation-induced cytidine deaminase gene (AID). The AID gene product increases somatic hypermutation of the p53 gene and promotes malignant transformation (based on Matsumoto et al. 2007).

Gastric cancer develops in 3% of individuals infected with *H. pylori*. Matsumoto et al. reported that p53 is mutated in 40% of cases of gastric cancer. MALT lymphoma, which develops in a small number of cases with *H. pylori* infection, is associated with four different specific chromosome translocations; three of these lead to activation of NF kappa B transcription factor (Sagaert et al. 2007).

Translational aspects: prevention of gastric cancer

Early detection of *H. pylori* infections and eradication of the organism, for example with clarithromycin, prevents the chronic infections that predispose to gastric cancer and MALT lymphoma. Clinical trials indicate that treatment is more successful in preventing gastric cancer in individuals who do not have atrophic gastritis (Kuipers and Sipponen 2006). Complete eradication of MALT lymphoma associated with cure of *H. pylori* infection was reported in 10 out of 15 patients by Fischbach (2002).

Kaposi Sarcoma: Activation of the G Protein–Coupled Receptor Gene *vGPCR*

Kaposi sarcoma is a vascular neoplasm that is a leading cause of mortality in AIDS (Montaner 2007). In certain African countries, including Uganda and Zimbabwe, it is the most common cancer in adult males. The etiologic agent is herpes virus HHV8, also known as Kaposi sarcoma herpes virus. One specific gene plays a critical role in causation of the sarcoma. This gene encodes a G protein–coupled receptor (vGPCR) that is expressed in dermal endothelial cells. This gene impacts the function of the $AKT/TSC/mTOR$ signaling cascade (see tuberous sclerosis, Chapter 7).

Kaposi sarcoma is rich in blood vessels. A number of intracellular cascades are activated by *vGPCR*. Among the most important is the *PI3K-AKT* pathway. *vGPCR* causes *AKT* activation, and this apparently protects endothelial cells from apoptosis. Manning and Cantley (2003) reported evidence that the hamartin tuberin complex, encoded by the tuberous sclerosis genes *TSC1* and *TSC2*, constitutes the link between the *PI3K-AKT* pathway and the *mTOR* pathway.

Sodhi et al. (2006) reported that *vGPCR*, through the *PI3K-AKT* pathway, promotes phosphorylation of the *TSC2* gene product tuberin, leading to its inactivation and its failure to act as Rheb GTPase. This in turn leads to phosphorylation and activation of *mTOR* by Rheb GTP and subsequently to activation of S6 kinase and eukaryotic translation initiation factor 4E binding protein 1. Inhibition of *mTOR* by rapamycin dramatically reduced proliferation of *vGPCR*-expressing cells.

Burkitt Lymphoma: DNA Damage and Chromosome Translocations

Virus-host cell interactions

Most individuals are silent carriers of EBV due in part to immune surveillance against virus-transformed, potentially malignant cells. EBV-induced lymphomas therefore occur in immune-suppressed hosts. In asymptomatic carriers, the expression of transforming proteins is downregulated. In silent infections, one virally encoded protein is primarily expressed.

EBV-immortalized lymphoblastoid cells express nine latency-associated viral proteins including the nuclear antigens EBNA 1-6. These latent viral proteins are known to activate the cell cycle and to inhibit apoptosis. However, EBV cell lines are nontumorigenic and retain the capacity for differentiation. Various patterns of viral expression occur and these are determined by virus-host cell interactions (Klein et al. 2007).

Burkitt lymphoma is a tumor primarily of children and it is endemic in parts of Africa and in New Guinea. In all endemic forms of Burkitt lymphoma, EBV is present. The key factor in EBV-induced malignancy is chromosome translocation that juxtaposes the *Myc* gene on chromosome 8 with one of the immunoglobulin loci on chromosome 14, 2, or 22. The type of immunoglobulin produced by the tumor cells is dependent upon the nature of the translocation, whether it involves heavy chain *Ig* loci (chromosome 14), kappa light chain *Ig* locus (chromosome 2), or lambda light chain (chromosome 22). Klein et al. (2007) reported that malignant transformation of *Ig-Myc* carrying translocations occurs in B cells that are producing high levels of cytokines due to chronic infections, such as endemic malaria infection or HIV infection. In addition, the B cells in Burkitt lymphoma do not express antigens that are recognizable by the *MHC1* (major histocompatibility locus 1) responsive cytotoxic lymphocytes.

Myc gene translocations are not specific for Burkitt lymphoma; they may occur in other lymphomas, such as large B cell lymphoma. However, in the latter forms of lymphoma the *Myc* locus is usually translocated to non-*Ig* loci. Hummel et al. (2006) reported that gene expression analysis using microarrays facilitates distinction of Burkitt and non-Burkitt lymphoma. This distinction is important because different chemotherapeutic protocols are required in these different forms of lymphoma. Burkitt lymphoma responds to high doses of cyclophosphamide and antimetabolic drugs. Using microarray analysis, Dave et al. (2006) reported that in Burkitt lymphoma there is high expression of a subset of lymphoid germinal center V cell genes and low expression of *HLA* class I genes.

Burkitt lymphoma cell lines may lose virus. Kamranvar et al. (2007) used a panel of EBV+ and EBV− Burkitt lymphoma cell lines to examine the effects of EBV on genomic stability. Chromosomal abnormalities in this condition comprise transmissible abnormalities and severe abnormalities that are nontransmissible. Genomic instability in Burkitt lymphoma is manifested by an increase in chromatid gaps, chromatid fragments, and dicentric chromosomes. These investigators used fluorescence in situ hybridization studies to examine telomeres in Burkitt lymphoma cell lines and determined that in EBV+ lines there is a high prevalence of telomere fusion, double-stranded breaks, and double-stranded break fusions. They concluded that DNA damage and telomere dysfunction are important mechanisms for EBV oncogenesis. They used specific antibodies to examine the phosphorylated variant of histone H2AX that occurs in regions adjacent to double-stranded DNA breaks. The presence of this variant indicates recruitment of DNA repair components. They determined that H2AX was present in higher quantities in EBV+ Burkitt lymphoma lines. EBV drives genomic instability and, in the absence of adequate immune mechanisms, malignancy results.

EBV infection may lead to lymphomas in other regions of the body. Lymphoma of the nasopharynx is common in parts of Asia. This tumor is frequently associated with rearrangement of T cell receptor genes.

Immune suppression and viral-related malignancies

The viral-related malignancies that EBV lymphoma and Kaposi sarcoma represent sometimes occur as complications of immunosuppression used following organ transplantation (Swinnen 2001). The increasing frequency of EBV-induced lymphoproliferative tumors in Western countries serves to stimulate the development of novel therapies. These include transfusion with allogeneic EBV-specific cytotoxic T cells (Haque et al. 2007).

EBV-induced DNA damage: role of AID and polymerase eta

Epeldegui et al. (2007) obtained important insight into the role of EBV in inducing tumors. They reported that infection with EBV resulted in increased expression of AID and polymerase eta. AID and polymerase eta are normally expressed in the germinal centers of lymphoid tissue and play a key role in the *Ig* heavy chain class switch recombination and somatic hypermutation. AID is not expressed in circulating B cells. It is expressed in embryonic stem cells. AID mutates C/G residues, and error-prone polymerase mutates A/T mutations. Both A/T and C/G mutations are common in somatic hypermutation. There is evidence that inappropriate expression of somatic hypermutation inducing molecules may play a role in the induction of lymphoma.

Epeldegui et al. published evidence that EBV-infected cells express increased amounts of AID and polymerase eta. They also demonstrated that EBV infection leads to progressive accumulation of mutations in the proto-oncogenes *BCL6* and p53. There is evidence that AID can induce mutations in actively transcribed genes. AID expression was also shown to be induced by hepatitis C, another lymphoma-inducing virus.

Polymerase eta is a translesion polymerase that plays a role in the repair of DNA damage. These polymerases carry out DNA synthesis across damaged bases and may use closely related sequences as a template for DNA repair. Polymerase eta plays a role in repair of double-stranded DNA breaks. Kawamoto et al. (2005) reported that polymerase eta plays an important role in the somatic hypermutation and gene conversion events associated determination of *Ig* diversity. Polymerase eta is deficient in mesodermal pigmentosum, a condition associated with defects in DNA repair and increased susceptibility to cancers, particularly of the skin. *Ig* somatic hypermutation is deficient in this condition. These investigators also demonstrated that cultured cells deficient in DNA polymerase etas show decreased *Ig* gene conversion.

Immunoglobulin (*Ig*) gene diversification requires alteration in gene sequence and in structure of the *Ig* genes. AID plays a key role in these processes (Vallur et al. 2007). In the process of somatic hypermutation, point mutations are introduced into rearranged and expressed *IgV* regions.

AID deaminates cytidine to uracil. Uracil is excised from DNA with uracil DNA glycosylase. The essential role of this enzyme in generating immunoglobulin diversification is evident in that absence of UDHG is associated with immunodeficiency syndrome. Uracil can undergo replication to T or it can be excised from the DNA strand.

A diagram of molecular mechanisms that play a role in malignant transformation is presented in Figure 9–2.

P53

Response to Genomic Damage and Imbalance

Tumor protein 53 (p53) has a variety of different functions including growth inhibition, tumor suppression, DNA damage response, and control of apo-

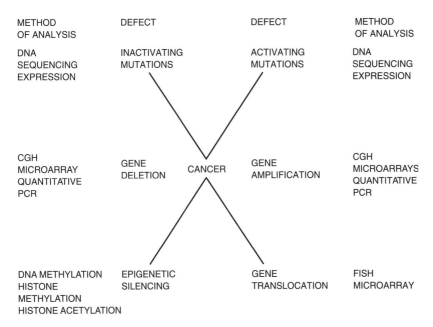

| METHOD OF ANALYSIS | DEFECT | | DEFECT | METHOD OF ANALYSIS |

Figure 9–2. Diagram of different genetic changes that lead to malignant transformation and cancer and methods used to detect these genetic changes (based on Haber and Settleman 2007).

ptosis. Increased production of p53 occurs in response to physiological stimuli, genomic damage, and imbalance in signaling pathways. Physiological factors that stimulate p53 production include metabolic stimuli, hypoxia, increased nitrous oxide production, and pH changes. DNA damaging agents (genotoxic stimuli) increase p53 production, promote cell cycle arrest, and in some cases stimulate apoptosis (Vogelstein and Kinzler 1992, 2004). Genotoxic agents such as X-rays induce p53 production; this in turn induces cyclin dependent kinase (CDK) inhibition, leading to cell cycle arrest.

DNA damage and p53

Double-stranded DNA breaks recruit ATM and ATR, proteins in the ataxia telangiectasia pathway. ATM signals ATR, and this in turn phosphorylates p53. A number of DNA damaging agents, including UV light and drugs, trigger activation of ATR and casein kinase and subsequently lead to p53 phosphorylation. Abnormalities in signal transduction, particularly those involving the Rb-E2F cycle (retinoblastoma-E2F transcription factor), may activate p53.

Tumor protein 53–induced apoptosis

Apoptosis induced by p53 serves to destroy potentially neoplastic cells. Through its inhibition of progress of the cell cycle by induction of p21cip1 that inhibits CDK and by induction of inhibitors of DNA polymerase, p53 prevents cells from entering S phase or progressing through the S phase. Indirectly, p53 affects expression of beta catenin, and this also impacts transition through the cell cycle. Other genes that act as targets and undergo increased transcription in response to p53 include growth arrest genes, DNA repair genes, apoptosis regulators, and antiangiogenesis proteins. Tumor protein 53 increases expression of the protein encoded by *MDM2* (a homolog of mouse double minute inducing gene). The product of this gene binds to p53, acts as a ubiquitin ligase, and promotes destruction of p53 in proteosomes. It plays a role in the transport of p53 from the nucleus to the cytoplasm. Through the feedback loop the cellular concentration of p53 is kept at low steady state levels. MDM2 is usually induced after DNA repair.

p53 Mutations and Altered Expression in Cancer

Mutations of p53 are present in approximately 50% of cancers in a variety of different tissues (Vogelstein et al. 2000). In other tumors where p53 is not mutated, its expression is affected indirectly by epigenetic mechanisms (Milyavsky et al. 2005).

The cancer-related mutations in p53 are usually missense mutations, and the mutant proteins are usually expressed and are full length. Furthermore, the mutant proteins are often expressed in high quantities. Some 97% of cancer-related p53 mutations occur in the DNA binding region. Mutations compromise the DNA binding capacity of p53 or the conformation of the protein. In some cases, p53 mutations have a dominant negative effect in that they compromise function of the wild-type p53. Weisz and colleagues (Weisz, Damalas, et al. 2007; Weisz, Oren, et al. 2007) reported that early in carcinogenesis, p53 mutations most often arise sporadically on one allele, and the mutant allele product may suppress functions of the wild-type allele product by forming oligomers with it. Later in the course of tumor progression, the remaining wild-type allele may be lost or mutated.

The mechanisms by which p53 mutations take place are generally poorly understood. The demonstration of increased expression of AID in response to NF kappa B and induction of p53 mutations through aberrant copying therefore represents an important breakthrough (Weisz, Damalas, et al. 2007).

Therapeutic Interventions Related to p53

Targeting protein folding

Many of the tumor-related p53 mutations lead to abnormal protein folding. In quantitative terms, the mutant DNA is often overexpressed in tumors. In many instances, the abnormal protein folding is not irreversible. A number of strategies are being implemented to use small molecules to target mutant p53 and alter its folding. In some cases, normal activity of mutant p53 has been achieved through use of antibodies that change the protein folding (Bykov et al. 2002; Selivanova and Wiman 2007).

Bykov et al. (2005) reported results of studies with a small molecule, PRIMA1, that restores wild-type conformation of p53 mutant proteins. PRIMA1 induced growth suppression in tumor cell lines and restored expression of genes that serve as downstream targets of p53. PRIMA1 induced apoptosis in a p53-dependent manner.

Strategies aimed at reducing p53 destruction by MDM2

Freedman and Levine (1999) reported that there are many possibilities for chemotherapeutic modulation of the interaction between MDM2-encoded protein and p53 (Figure 9–3). Vassilev et al. (2004) noted that analysis of the crystal structure of MDM2 protein revealed that it has a deep hydrophobic pocket that offers potential for interaction with small molecules. They identified a series of cis-imadazoline analogs named nutlins that displace p53 from its complex with MDM2 protein. They demonstrated that treatment of tumor cells with these compounds led to elevated levels of p53 and promotion of the downstream effects of p53, including cell cycle arrest. One difficulty with the use of these agents in systemic therapy is that their effects are not restricted to tumor tissue. In further studies of nutlins, Stuhmer et al. (2005) reported that MDM2 antagonists do not have genotoxic effects. They concluded that nutlin antitumor activity might be achieved without the side effects of other chemotherapeutic agents.

Other small molecules under investigation are designed to reduce the ubiquitin ligase activity of MDM2 toward p53 (Buolamwini et al. 2005).

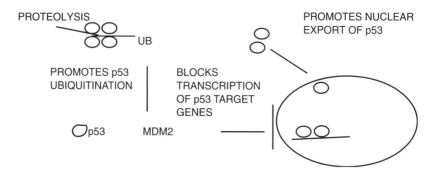

Figure 9–3. MDM2 acts as a negative regulator of p53 by blocking transcription of p53 target genes, promoting nuclear export of p53, and expediting ubiquitination of p53 and its proteolysis. A number of therapeutic agents used to treat cancer are compounds that inhibit MDM2 (based on Vassilev et al. 2004).

DESIGNING IDEAL CANCER THERAPY

Anticancer Drugs That Impact Single Targets

Based on molecular genetic studies in tumor cells, a number of specific therapeutic agents have been developed that target tumor cell receptors or specific proteins in signal transduction pathways. These include small molecule kinase inhibitors and therapeutic antibodies. Molecular and structural analysis of specific proteins, including analysis of crystallographic structure and definition of functional domains, provides basic information that may contribute to drug design. Small molecule kinase inhibitors in clinical use include imatinib (Gleevec), gefitinib (Iressa), and erlotinib (Tarceva). Therapeutic antibodies used include trastuzumab (Herceptin), cetuximab (Erbatix), and bevacizumab (Avastin).

Multiple Targets: Molecular Chaperone HSP 90

Powers and Workman (2006) reported that multiple signaling pathways are often disrupted in cancer, and this limits the therapeutic potential of drugs that impact a single target. They noted further that cancers due to mutations in a single gene might develop resistance due to ongoing mutations. They emphasized the importance of investigating agents that impact multiple signaling pathways, for example, chaperones that impact activity of heat shock protein, HSP 90.

Oncogenic signaling	Angiogenesis promoters	Receptors	Replication promoters
BCR-ABL	HIFalpha	Androgen receptor	TERT
AKT-PKB	VEGF	Estrogen receptor	(Telomeres)
RAF-MET			
Mutant p53			

Figure 9– 4. Heat shock protein HSP 90 acts as a molecular chaperone for different classes of tumor-promoting proteins (based on Powers and Workman 2006; Sharp et al. 2007). Inhibitors of HSP 90 are being developed to treat cancer.

Molecular chaperone HSP 90 plays a role in determining the stability of a number of oncogenic proteins including p53, ERBB2, BRAF, CRAF, CDK, steroid hormone receptors protein kinase (*AKT*), and other protein kinases and telomerase (Sharp et al. 2007; Figure 9– 4). Inhibition of HSP chaperone proteins leads to degradation in the ubiquitin proteosome pathway of proteins to which they are bound. Eukaryotes require heat shock proteins for viability.

HSP 90 often undergoes association with cochaperones prior to its interaction with the target protein. For example, association of HSP 90 with steroid hormone receptor involves a prior interaction with HSP 70 and HSP 40 and an additional cochaperone.

The N terminal domain of HSP 90 has an ATP binding site. Inhibitors of HSP 90 function are competitive inhibitors of ATP binding. HSP 90 has low inherent ATPase activity. A key function of HSP 90 is hydrolysis of ATP and ADP-ATP exchange. Inhibition of HSP 90 function may be achieved by interruption of its ATPase activity. The middle section of the protein acts as a client-binding domain. The mechanism through which HSP 90 ATP binding and ATP hydrolysis to ADP affect the client protein is still under investigation.

One drug that is in clinical trials as an HSP 90 inhibitor is the geldanamycin analog 17-allylamino-17-demethoxygeldanamycin. This drug is used as a single chemotherapeutic agent or to enhance efficacy of other chemotherapeutic agents. Other HSP inhibitors are under development, since a side effect of geldanamycin is hepatotoxicity.

CANCER STEM CELLS

There is evidence that a small population of cells, called cancer stem cells, primarily determines tumor growth. Stem cell theory is based on the observation that the vast majority of cells within a tumor do not have the capacity to give rise to tumors on transplantation. Methods for identifying tumor cells include transplantation into immune-deficient mice (NOD/SCID mice). Cancer stem cells may be defined as self-renewing, multipotent tumor-initiating cells.

Further indications of the importance of stem cells in cancer initiation comes from evidence that cancer stem cells have many properties in common with normal stem cells in that same tissue. Furthermore, in the hematopoietic system, for example, only stem cells are sufficiently long lived to accumulate the mutations necessary for malignant transformation.

If cancer stem cells are the main cells involved in tumor development, it will be critical to develop therapies that specifically target these cells. It is, however, important to consider the relationship of cancer stem cells to normal stem cells for specific tissue. Close relationships between normal and cancer stem cells may hinder development of stem cell-specific therapies (Yilmaz et al. 2006). There is, however, progress in identifying properties that distinguish between these two stem cell types. Furthermore, there is progress in development of specific therapies that target the differences.

One difference between normal hematopoietic stem cells and leukemic stem cells is the expression in normal stem cells of PTEN (phosphatase and tensin homologue). Yilmaz et al. (2006) and Zhang et al. (2006) reported that PTEN plays an important role in stem cell activation, differentiation, and lineage determination. Yilmaz et al. reported that PTEN depletion in mice results first in myeloproliferative disease, and within weeks leukemia occurs in these mice. Loss of PTEN leads to activation of *mTOR* through its effect on *AKT* kinase. Rapamycin treatment of PTEN$-/-$ mice reduced leukemic stem cells and also restored the normal bone marrow stem cells. This observation is important since many chemotherapeutic agents used in the treatment of leukemia or lymphoma are often not effective in killing off self-renewing cancer stem cells.

Through its effect on *AKT*, PTEN also indirectly influences FOXO (forkhead box O) transcription factors. These factors play a critical role in the transcription of genes that encode products that play an important role in stem cell maintenance (Trotman et al. 2006).

STEM CELL NICHES

Calabrese et al. (2007) proposed that brain cancer stem cells are maintained in a specific vascular niche and that vascular endothelial cells secrete factors that maintain cells as stem cells. They demonstrated that increases in endothelial cells or blood vessels are associated with accelerated production of cancer stem cells. They therefore proposed that antiangiogenic drugs (such as becacizumab) can indirectly inhibit stem cell growth.

Becacizumab is a monoclonal antibody against vascular endothelial growth factor. It has been shown to be advantageous in treatment of breast cancer, particularly when used in combination with other chemotherapeutic agents.

Neural stem cells and brain cancer cells express nestin, an intermediate filament protein, and are positive for the CD133 antigen (prominin1). Calabrese et al. (2007) reported that in brain tumors, including medulloblastoma, ependymoma, oligodendroglioma, and glioblastoma, stem cells are located in proximity to tumor capillaries. They determined that the number of nestin-producing cells overall is low in tumors. However, the numbers of nestin-positive cells is significantly higher in proximity to tumor capillaries. Their studies on cultured tumor stem cell masses revealed that growth of these masses was greatly enhanced in the presence of primary endothelial cells. These studies led them to conclude that endothelial cells likely secrete soluble factors that maintain stem cells and that angiogenic signaling is critical for growth. They demonstrated further that becacizumab, which is active against VEGF (vascular endothelial growth factor), depleted blood vessels and disrupted stem cell self-renewal in models of medulloblastoma and glioma. There are favorable reports on clinical trials where bevacizumab was used in combination with the chemotherapeutic drug CPT11 in the treatment of glioblastoma (Reardon and Wen 2006).

HYPOXIA IN TUMOR MASSES AND HYPOXIA-INDUCIBLE FACTORS

Cellular proliferation in cancer leads to the formation of masses that are poor in nutrients and low in oxygen. These conditions have a number of consequences, one of which is increased production of hypoxia-inducible factors (HIF). Another factor shown to induce HIF is loss of tumor suppressor genes. Maxwell and Salnikow (2005) reported that HIF1-alpha subunits are present in the majority of tumors and that their concentration differs considerably in different parts of the tumor. This is also true for concentrations of VEGF in tumors. High levels of HIF production in tumors are associated with a poor prognosis. HIF activates transcription of a number of different genes and leads to a number of changes including increased angiogenesis and erythropoiesis, increased glycolysis and glucose transport, and increased telomerase production.

Biological Activities of HIF

Hypoxia-induced factors are a family of transcription factors that play a key role in the response of mammalian cells to oxygen deprivation. HIF1 is composed of two subunits, alpha and beta. HIF1 was first discovered as an inducer of erythropoietin production by renal cells in response to hypoxia. The HIF1-alpha subunit mediates cellular response to hypoxia. There are three domains in the HIF1-alpha subunit that have different functions. The N terminal domain determines DNA binding. The C terminal domain is important in ensuring nuclear localization of HIF1-alpha. The central portion of HIF1-alpha plays a key role in oxygen response; within this region there is an oxygen-dependent degradation domain; at each end of this domain there are proline residues. Prolylhydroxylases act as oxygen sensors. In the presence of normal oxygen tension, HIF1-alpha proline residues are hydroxylated. This requires the enzyme proline hydroxylase and takes place in the presence of oxygen (O_2) and iron (Fe^{++}). Proline hydroxylation facilitates HIF1-alpha reaction with the von Hippel Lindau protein (VHL). This interaction targets HIF1-alpha for polyubiquitination and degradation in proteosomes. There is evidence that the interaction of HIF1-alpha with VHL is further facilitated by a protein ARD1 (arrest defective protein 1) that facilitates acetylation of lysine residues.

The activity of HIF1-alpha is also impacted by hydroxylation of asparagine residues through activity of asparagine hydroxylase. Under conditions of normal oxygen tension, asparagine hydroxylation prevents HIF1-alpha from recruiting a transcriptional coactivation protein, p300CBP.

Steady-state levels of HIF1-alpha are low. Under hypoxic conditions, HIF1-alpha accumulates. It is stabilized and it is phosphorylated by *MAPK* (mitogen-activated kinase). This interaction increases HIF1-alpha transcriptional activity, likely through increasing interaction with the HIF1-beta subunit and the p300CBP coactivator.

Modulation of HIF1-alpha Expression in Tumors

Loss of expression of p53 tumor suppressor leads to increased expression of HIF1-alpha. Activation of signaling pathways via tyrosine kinase receptors, for example, through mitogen-activated kinase (*MEK*) and phosphatidyl-inositol-3-kinase (*PI3K*), leads to activation of *AKT* and induces HIF1-alpha expression. Loss of PTEN further enhances HIF1-alpha production.

Hypoxia and HIF1 play an important role in induction of VEGF. HIF1 also induces expression of nitric oxide synthase that further influences tumor vascularization.

Under conditions of low oxygen concentration such as exist in rapidly growing tumors, cells employ anaerobic glycolysis, and HIF1-alpha plays a key role in this metabolic shift. Under aerobic conditions, pyruvate enters mitochondria and is converted to acetyl-CoA. In red cells where mitochondria are absent and under aerobic conditions, reoxidation of NADH to NAD through the mitochondrial respiratory chain does not occur. Generation of NAD through NADH occurs through conversion of pyruvate to lactate through the activity of lactate dehydrogenase. The NAD thus generated is used to promote glycolysis.

The redox reaction whereby pyruvate is converted to lactate is carried out by lactate dehydrogenase A (LDHA) and regenerates NAD+, which acts as an electron acceptor (Fantin et al. 2006).

HIF1 induces expression of a number of enzymes in the glycolytic pathway including aldolase, phosphoglycerate kinase, and lactate dehydrogenase. It also induces expression of glucose transporter, Glut1. In addition, hexokinase is induced so that glucose taken up by the cells is transformed to glucose-6-phosphate.

HIF1 and Telomeres

HIF1 induces expression of hTERT, the catalytic subunit of telomerase. In tumor cells characterized by DNA damage and uncontrolled proliferation, telomeres are progressively shortened. This process and the resulting end fusions lead to cell death. By stimulating telomerase activity (and restoring telomeres), HIF1 rescues cells destined for senescence and contributes to immortalization (Kang and Park 2007).

Therapeutic Implications: Targeting HIF

There are a growing number of therapeutic strategies targeting HIF. Blocking HIF, for example with small molecules, makes tumors more susceptible to treatment with chemotherapeutic agents. Hypoxia in tumors has a marked effect on response to therapy. Resistance to chemotherapeutic agents and radiation are common features of hypoxic cells and tumors (Melillo 2007). Drug distribution is impacted in part through increased interstitial pressure. There is also evidence that hypoxic cells are more invasive and metastatic. Majumder et al. (2004) reviewed therapies that target HIF1. There is evidence that HIF1-alpha activity is reversed by *mTOR* inhibition with rapamaycin. The precise mechanism through which this occurs is not clear.

HIF1-alpha activity is decreased by treatment with 2-methoxyestradiol. This drug is beneficial in treatment of carcinomas of the head and neck (Ricker et al. 2004).

Geldanamycin, known to inhibit the chaperone protein HSP 90, destabilizes HIF1-alpha and induces its degradation. Erythropoietin is secreted by renal fibroblasts in response to hypoxia and its expression is increased by HIF1. Increasingly, erythropoietin is added to chemotherapy protocols to increase the patient's hemoglobin concentration and reduce hypoxia.

An active metabolite of the topoisomerase inhibitor irinotecan inhibits endothelial cell proliferation and decreases expression of HIF1-alpha and VEGF.

METABOLISM IN TUMORS

Warburg in 1930 first reported that anaerobic metabolism of glucose predominates in tumors. Shaw (2006) proposed that alteration in metabolism plays an integral role in tumorigenesis. He noted that positron emission tomography scan with fluorodeoxyglucose revealed that glucose metabolism is altered in the majority of tumors. Key factors in the increased glucose uptake in tumors and increased anaerobic glycolysis are increased activity of HIF1 and activation of *AMPK* (adenosine monophosphate activated protein kinase). HIF1 downregulates oxidative phosphorylation and the TCA cycle through promoting the activity of pyruvate dehydrogenase kinase. Pyruvate dehydrogenase is inhibited, allowing pyruvate to accumulate. As described above, pyruvate is converted to lactate through activity of lactate dehydrogenase, allowing regeneration of NAD, which facilitates further anaerobic glycolysis.

There is increasing evidence that the switch from mitochondrial energy generation to anaerobic glycolysis provides an advantage to tumor cells. Loss of mitochondrial respiration leads to activation of the *PI3K-AKT* pathway and provides tumors with a growth advantage (Fantin and Leder 2006).

AMPK is activated by a kinase, designated *LKB1*, identified as deficient in Peutz-Jeghers syndrome. Following activation, *AMPK* phosphorylates a number of downstream targets. It phosphorylates sites on tuberin that are distinct from sites phosphorylated by *AKT*. *AMPK* activation of the *TSC1-TSC2* complex leads to inhibition of *mTOR*. Deficiency of *TSC*, PTEN, or *LKB1* leads to hyperactivity of *mTOR* that facilitates tumor growth. There is also evidence that *AMPK* phosphorylates and activates p53 (Feng et al. 2007).

AMPK is activated by a number of agents used in the treatment of diabetes mellitus, such as metaformin and thiazolides. These drugs may be useful in treatment of tumors (Motoshima et al. 2007).

REACTIVE OXYGEN SPECIES AND CANCER

Reactive oxygen species are chemical species with one or more unpaired electrons. Superoxide is a free radical of oxygen that has an unpaired electron, O_2^-. It is unstable and converts to peroxide O_2^{2-} with two free electrons and to H_2O_2 (hydrogen peroxide). ROS generated in mitochondria under hypoxic conditions create a flux of H_2O_2 that inhibits prolyl hydroxylase and increases HIF1-alpha concentrations (Fruehauf and Meyskens 2007). ROS are also increased in tumors due to oncogene signaling via NADPH oxidase. Activation of *PI3K* and inositol 1,4,5, triphosphates activate expression of the proto-oncogene Rac, which in turn activates NADPH oxidase, leading to H_2O_2 production. Membrane-bound NADPH oxidase plays a key role in the generation of ROS. ROS oxidize catalytic cysteines in protein tyrosine phosphatases and in protein kinase C. H_2O_2 thus inhibits protein phosphatases.

Growth factors and cytokines may induce production of ROS, either through NADPH oxidase or in a mitochondrial dependent manner (Wu 2006). There is evidence for increased production of ROS in tumors and evidence that ROS play a role in tumor progression (Wu 2006). The role of ROS in determining response to cancer therapy is the subject of intense investigation (Gius and Spitz 2006).

MOLECULAR GENETIC ANALYSIS OF COLON CANCER AND RELEVANCE TO THERAPY

The most important known factors predisposing to colon cancer are the genetic conditions hereditary nonpolyposis coli and familial adenomatous polyposis (FAP) and the inflammatory condition ulcerative colitis (Clevers 2006).

Genetic Predisposition

The *APC* gene (adenomatous polyposis coli) is mutated in FAP and in a high proportion of cases of sporadic colorectal tumors. FAP is characterized by the presence of hundreds or thousands of colorectal adenomas, one or more of which often become malignant by the fourth decade. *APC* gene mutations that cause FAP most frequently lead to synthesis of a truncated protein. The most common mutations are C to T (cytosine to thymine), resulting from spontaneous deamination of 5-methylcytosine (Segditsas and Tomlinson 2006). Large deletion mutations of *APC* occur in some cases of FAP. Loss of heterozygosity studies revealed that a second hit at the wild-type allele is necessary for tumor formation. The second hit deletion often involves regions of the *APC* gene that interact with beta-catenin in the *Wnt* pathway (see p. 149).

Fewer than 5% of cases of colorectal cancer are due to high-penetrance inherited familial conditions, such as FAP, Lynch syndrome (familial p53 mutations), or mismatch repair gene defects. Somatic mutations occur in the majority of cases.

Somatic Mutations

Sjoblom et al. (2006) analyzed somatic mutations in colon and breast cancer tumors (11 of each) through comprehensive sequence analysis of 13,023 genes. They concentrated on sequences within consensus protein coding regions of each gene. They discovered that each tumor exhibited approximately 90 mutations; these were primarily missense mutations. This likely attests to the ongoing mutation process that occurs in malignant tumors. Haber and Settleman (2007) noted that by the time a cancer is diagnosed, it carries cells that not only have the abnormality that initiated malignant proliferation but also additional genetic changes. Some mutations may therefore be considered as drivers while others are incidental passengers.

Inflammatory Processes in Colon Cancer and NSAIDs

Ulcerative colitis increases the risk of colon cancer by an order of magnitude. Clevers (2006) noted that there is additional evidence for the role of inflammatory processes in the etiology of colorectal cancer. Nonsteroidal anti-inflammatory drugs (NSAIDs) reduce the death rates in sporadic colorectal cancer; these drugs reduce prostaglandin synthesis. He notes further that NSAID therapy leads to regression in adenoma size in patients with FAP. Of further interest is the observation that NSAIDs and other inhibitors of COX enzymes (cyclooxygenases) reduce the risk of adenomas in mice with mutation for *APC* genes. Furthermore, if a second mutation, involving the prostaglandin receptor gene *EP2*, is introduced into the *APC* mutant mice, the risk of adenomas is decreased.

Castellone et al. (2005) elucidated one pathway through which inflammatory prostaglandins impact colon cancer risk. Key elements in this pathway include WNT (Wingless protein), beta-catenin transcription factor, tumor suppressor molecules APC and axin, and the serine kinase GSK3B (glycogen synthase kinase 3B). Extracellular-derived WNT molecules bind to cell surface receptors. This binding inhibits the axin-APC-GSK3B complex, and beta-catenin is stabilized. Beta-catenin rises and enters the nucleus where it acts as a transcription factor for specific target genes that play a role in cell proliferation. When WNT signaling is absent, the axin-APC-GSK3B complex is active, and this degrades beta-catenin. The downstream effect is that beta-catenin target genes are not transcribed. *APC* mutations impact the function of the complex and stabilize beta-catenin.

Castellone et al. reported that prostaglandin E2 (PGE2) binds to its receptor, and this binding activates the cytoplasmic G-coupled receptor, which in turn activates the APC-axin-GSK3B complex that phosphorylates beta-catenin. Phosphorylated beta-catenin enters the nucleus and stimulates transcription. NSAIDs interrupt this process, and the downstream effect is dissociation of the APC-axin-GSK3B complex and process and catenin degradation.

Markowitz (2007) reviewed the upstream pathway through which NSAIDs influence prostaglandin synthesis and lower the incidence of colon cancer. Prostaglandin is synthesized from arachidonic acid through a pathway that requires a series of enzymes including Cox1, Cox2, PGH2 (prostaglandin H2), and prostaglandin synthase. NSAIDs impact Cox2 activity. Interestingly, Cox2 is present in increased amounts in colon cancer. This increase is likely due to decreased activity of the enzyme 15PGDH (15-hydroxyprostaglandin dehydrogenase). There is evidence from animal studies that 15PGDH inhibits the development of intestinal neoplasias.

Markowitz (2007) noted that PGE2 is particularly important in the growth of colon cancer cells. PGE2 binds to the prostaglandin receptors EP1, EP2, EP3, and EP4. This receptor-ligand interaction activates signaling pathways, especially *PI3K*, beta-catenin transcriptional activity, and expression of PPAR gamma (peroxisomal proliferators activated receptor gamma). The downstream effects of these activities lead to increased production of cyclins, BCL2, and VEGF, proteins involved in growth, migration, antiapoptosis, and angiogenesis respectively.

Chan et al. (2007) reported that Cox2-positive colorectal tumors occurred less frequently in regular users of aspirin. They noted further that 67% of colorectal cancer tumors exhibit moderate or strong Cox2 expression.

Markowitz (2007) and others have emphasized that aspirin use has significant side effects and that alternate strategies should be sought to inhibit this pathway. These strategies may include inhibitors of PGE2 receptors or of PGE2 synthase. Inhibitors of downstream targets may also be considered for therapy. These include inhibitors *PI3K*, PPAR gamma, or 15PGDH.

Markowitz noted that a further challenge involves identification of individuals with colorectal tumors in which *COX2* is overexpressed.

Other Frequent Mutations in Colorectal Cancer

In addition to *APC* mutations or deletions, mutations in the K-*ras* and p53 genes are often present in colorectal cancers, though K-*ras* and p53 mutations seldom occur in the same tumor. Conlin et al. (2005) reviewed the effects of mutations in these genes and reported that patients with K-*ras* mutations had a significantly poorer prognosis than patients without mutation in that gene.

As described above, the *APC* gene plays a role in the WNT beta-catenin signaling pathway and its downstream effects impact cell cycle regulation and apoptosis and, in addition, intercellular adhesion and cytoskeletal stabilization.

The DNA binding transcription factor p53 impacts the cell cycle and specifically causes delays in cell cycle progression to repair DNA damage. In some cases, p53 promotes apoptosis of cells with damaged DNA.

K-*ras* is a small molecule involved in signal transduction; it has intrinsic GTPase activity. The K-*ras* mutations found in colon cancer often abolish GTPase activity and thereby promote proliferation. In a study of 107 cases, Conlin et al. (2005) determined that *APC* mutations occurred in 56%, K-*ras* in 27%, and p53 in 61% of tumors. Mutations involving all three genes occurred in 6% of tumors. They noted that K-*ras* mutations occurred particularly in late-stage tumors. Because of the importance of K-*ras* mutations, extensive studies are in progress to target K-*ras*. These studies include RNA interference strategies (see p. 151).

It is important to note that K-*ras* is mutated in 90% of pancreatic cancers and that currently treatment of pancreatic cancer is very unsatisfactory.

Targeting Aberrant Metabolism in Colon Cancer

Gerner et al. (2007) reported that downstream effects of *APC* mutations in colon cancer include increased expression of the enzyme ornithine decarboxylase, and that this leads to the presence of increased concentrations of polyamines. They noted that the *APC* mutations in individuals with familial adenomatous polyposis lead to increased polyamine concentrations in precancerous colonic tissue. These authors noted further that altered transcription that results from *APC* mutations, and particularly altered transcription of PPARG, impacts catabolism of polyamines. They reported successful phase 2 clinical trials that involved the use of an NSAID and a compound, difluoromethylornithine, that stimulates polyamine acetylation and transport.

RNA INHIBITION AND ANTISENSE OLIGONUCLEOTIDES IN CANCER THERAPY

In vitro studies have demonstrated the efficacy of RNA inhibition (RNAi) to silence specific genes involved in tumor progression and to promote apoptosis (Pai et al. 2006). An important role for RNAi technologies is to knock down expression of specific genes in in vitro systems or in mouse models of a specific tumor. These technologies facilitate identification of genes that promote tumor growth and of genes that function in interacting pathways.

Obstacles to in vivo siRNA (small inhibitor RNA) therapy include delivery systems, difficulties with nonspecific immune responses, effects on expression of related genes, and incomplete suppression of the target gene. Coupling of siRNA to nanoparticles or liposomes facilitates delivery of siRNA. There are reports that specific modifications of siRNA may improve stability; these include modifications to the ribose moieties. Special delivery systems may facilitate targeting to specific tissue; these include use of antibodies.

RNAi technologies may potentially be used in cases where oncogenes are overexpressed due to dominant mutations. Pai et al. (2006) emphasized the potential importance of RNAi applications when overexpression of particular genes leads to drug resistance or radiation insensitivity of tumors.

In addition to silencing oncogenes, RNAi technologies can be used to impact genes that negatively regulate tumor suppressor genes. Antiapoptotic proteins play an important role in certain cancers, and these genes may potentially be silenced by RNAi. Another potential application for RNAi will be to block tumor angiogenesis. Triplex-forming oligonucleotides to target and inactivate the promoter region of oncogene *cMyc* were described by Christensen et al. (2006). Conjugation of these oligonucleotides to anticancer nucleosides, for example, gemcitabine or daunomycin, generated specifically targeted molecules that proved effective in reducing cancer cells in experiments in vitro.

NEW APPROACHES TO CANCER TREATMENT: FACTORS THAT INDUCE REVERSION TO NORMAL

The selection of molecules for investigation as possible chemotherapeutic targets against cancer is often based on identification of differences between cancer and normal cells. Tuynder et al. (2004) utilized a different approach and sought to identify molecular changes that caused cancer cells to revert to normal. Their studies were based on earlier work by Todaro et al. (1980), who first described the phenomenon of reversion in a mouse tumor cell line, NIH3T3. Revertants regained their sensitivity to contact inhibition in cell culture.

Tuynder et al. (2004) studied cell lines developed from human tumors including leukemia cell and solid tumors from colon and lung and melanoma cell lines. They identified a protein, TCTP (translationally controlled tumor protein), that is downregulated in tumor reversion.

In separate experiments, they determined that downregulation of TCTP with antisense RNA promoted apoptosis. TCTP has properties of a histamine-releasing factor. Tuynder et al. (2004) demonstrated inhibition of TCTP effects following administration of certain types of antihistamines.

Structural studies have demonstrated that TCTP has homology to GTP/GDP binding proteins and that it interacts functionally with translation elongation factor eEF1A and its guanine nucleotide exchange factors.

Bommer and Thiele (2004) reported that TCTP levels are regulated in response to extracellular signals and that high TCTP levels occur in mitotically active tissues. Interestingly, TCTP is highly expressed in *Plasmodium* and in parasitic worms. The parasite-derived TCTP induces an inflammatory response and histamine release in the host. *Plasmodium* TCTP binds to the antimalarial drug artemisinin. Increased expression of TCTP is associated with increased resistance to this drug. Artemisinin and its derivatives also have potent antitumor activity. Artemisinin is derived from a plant used in Chinese medicine, sweet wormwood, *Artemisia annua qinghao*. Artemisinin analogues are being investigated as anticancer drugs.

SUMMARY: MOLECULAR GENETICS-BASED TREATMENT OF CANCER

Cancer treatments in the era of molecular genetics and genomics build on research that identifies molecular targets for therapy. Weinberg (2007) reported that some of the most effective drugs in cancer treatment are drugs that have an enzymatic function, target an enzyme or a protein that has a catalytic cleft or a drug-binding pocket, or are small molecules that penetrate to the interior of tumors. Many drugs under investigation are designed to target kinases that are abnormally active in tumors.

Through use of tumor profiling or stratifying and analysis of gene expression in tumor cells through array technology, researchers claim to have developed a high degree of accuracy in determining which tumors require aggressive therapy. This is particularly the case for breast cancers and certain lymphomas (Weinberg 2007). It is important to note that tumors with very similar histological appearance may have very different long-term survival rates. In the long-term expression, arrays may be particularly useful in designing therapies.

The goal of specific molecular targeting is to eliminate toxicity to normal tissues. There is increasing evidence that many different genetic aberrations occur in malignant tumors. Genomic instability is a hallmark of malignant tumors, and this is manifested by the presence of mutations in many different genes and by chromosome gains and losses. The question then arises whether therapies that target a particular molecular genetic change are useful in treatment. Fletcher (2004) reviewed molecular targeting in cancer treatment. He noted that although many different genetic events occur in tumors, there is often one rate-limiting or initiating event, and frequently a specific abnormal gene product plays a key role in tumor development. He noted further that frequently the specific behavior of a tumor may be predicted by the unique constellation of genetic changes and the pattern of gene expression. Expression of a particular oncogene may, for example, be indispensable for tumor growth.

10

PHARMACOGENETICS AND PHARMACOGENOMICS

Pharmacogenetics and *pharmacogenomics* are terms that are often used interchangeably. However, the National Center for Biotechnology Information (NCBI) defines them differently: "Pharmacogenomics may refer to the general study of genes that determine drug behavior. Pharmacogenetics is the study of inherited variation in drug metabolism and response" (Figure 10–1).

CHEMOGENOMICS AND DRUG DESIGN

Recent developments in genomics have led to significant changes in strategies for drug design. Rognan (2007) reviewed chemogenomic strategies and noted that changes are based on the following:

- The availability of genome sequences of at least 180 different organisms
- High-throughput miniaturization of chemical synthesis
- The potential for biological evaluation of multiple compounds and their effects on gene and protein expression

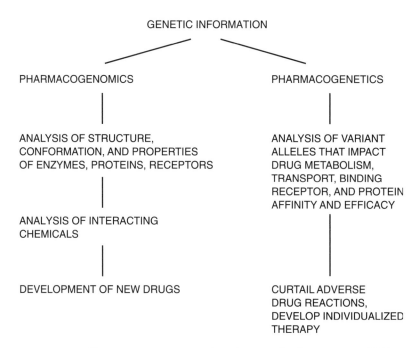

Figure 10-1. Pharmacogenomics involves the use of information on DNA sequence, protein sequence, structure, conformation, and function to develop new drugs. Pharmacogenetic studies investigate variant alleles that alter drug metabolism, transport, target affinity, and efficacy and are essential in prevention of adverse drug reactions and development of approaches to individualized drug therapy.

- Possibilities for investigation of drug effects on families of related proteins
- Possibilities for investigation of effects on full metabolic pathways

Chemogenomics refers to analysis of the biological effects of small ligands on macromolecular targets. Caron et al. (2001) reported on the potential for expanded drug discovery through miniaturization and of chemical synthesis, high-throughput in vitro biological screening, and computational methods to determine structure and activity relationships. In silico approaches are often used in drug design and are essential in analysis of information on drug action. Paolini et al. (2006) reported that by 2006 the pharmaceutical industry had only examined approximately 3000 of the 25,000 products encoded by the human genome; many more targets need to be analyzed.

A great leap forward was made in medicinal chemistry through miniaturization of synthesis and parallelization of synthesis of compounds. Potential compounds that may be synthesized are sometimes referred to as the chemical space. This includes more than 10 million nonredundant chemical structures (Rognan 2007). The library of compounds available includes potential ligands for gene products. The goal of chemogenomics is to match ligands and targets. In silico analyses may be used to predict interactions. This analysis is predicated on the assumption that compounds of similar chemical structure will interact with the same target. Furthermore, targets that are related by virtue of their sequence and structure may react with the same ligand. Target proteins are classified according to their sequence and structure. The two-dimensional structure includes analyses of alpha helices and beta sheets or coils. The three-dimensional structure is determined through analysis of atomic coordinates derived from X-ray diffraction, nuclear magnetic resonance imaging, or molecular modeling. Chemicals (potential ligands) are described by properties including molecular weight, spectroscopic bands, topological features, and one-dimensional, three-dimensional properties. Separate chemical and biological databases may be merged to generate a chemogenomic database.

Paolini et al. (2006) reviewed the concept of developing drug discovery into a knowledge-based predictive science that incorporates information on medicinal chemistry and information on protein sequence. They developed an integrated database that by 2006 included 4.8 million nonredundant chemical structures, 275,000 of which were biologically active. They referred to this as the pharmacological space.

In an interesting application of chemogenomics and structural analysis to design of drugs for cancer treatment, Hurley et al. (2006) described the presence of a parallel-stranded G nucleotide quadruplex in the purine-rich region of the promoter of the *cMyc* oncogene. This quadruplex structure acts as a silencer element, and its structure is stabilized by the cationic porphyrin TMPYP4. Hurley et al. noted that this cationic porphyrin represses *cMyc* expression. They note further that similar G nucleotide–rich quadruplex structures occur in telomere regions and in a number of other oncogenes. These observations hold promise for the development of antioncogene drugs that are not based on enzymatic activity.

PHARMACOGENETICS

Adverse Drug Reactions

Adverse drug reactions are responsible for at least 100,000 deaths per year in the United States (Lazarou et al. 1998). In addition to general toxicity, in susceptible individuals a specific drug or class of drugs may damage a particular organ or system, for example, aminoglycoside-induced deafness.

There are also individual differences in the effectiveness of drugs. Specific drugs are almost useless in patients with a particular genetic makeup. These include genetic variations (polymorphisms) that occur in proteins responsible for the following:

- Transport and tissue targeting of drugs
- Activation of drugs
- Removal or detoxification of drugs
- Receptors for drugs
- Passage of drugs into cells via channels or via transport proteins
- Immune system gene polymorphisms that play a role in certain drug tolerance problems and drug allergies

Adverse Drug Reactions Due to Defective Detoxification of Drugs

Polymorphisms in cytochrome p450 enzymes (CYP) account for 80% of adverse drug reactions. Since these enzymes are responsible for metabolism of a variety of drugs, a specific CYP polymorphism may lead to abnormal sensitivity to a number of different drugs. Polymorphisms in CYP2D6 are particularly common and important.

Before prescribing medications, physicians are advised to check whether there is evidence in the literature that the drug is known to be metabolized by a polymorphic allele. The prevalence of specific polymorphic alleles of the relevant drug-metabolizing enzyme varies in different populations so that the ethnicity of the patient being treated must be taken into account (Phillips et al. 2001). Physicians must also be aware of the possible hazards of prescribing concurrently two or more drugs that are known to be metabolized by the same polymorphic enzyme. In cases where adverse drug reactions occur, studies should be undertaken to define the cause of the adverse reaction and to determine if it is due to a genetic variant.

Adverse drug reactions, such as neurological side effects of the antituberculosis drug isoniazid, may also be due to polymorphisms in *N*-acetyltransferase (NAT). The uridine diphosphoglucuronosyltransferase (UGTs) are encoded by a family of genes. Thus far, 17 different human UGT transcripts have been identified (Desai et al. 2003). A variety of drugs, xenobiotics, and endobiotics are conjugated by enzymes of the UGT1A and UGT1B families. UGT polymorphisms have now been documented. Several polymorphisms alter enzyme activity and their role in altered xenobiotic metabolism is being investigated (Nagar and Remmel 2006). There is evidence that a common deletion polymorphism on chromosome 4q13 leads to deletion of UGT2B17 UDP glucuronosyltransferase (Wilson et al. 2004). This enzyme metabolizes androgens. This deletion polymorphism is associated with elevated levels of endogenous androgens (Jakobsson et al. 2006). This polymorphism, through its effect on androgens, may also influence the risk of prostate cancer (Park et al. 2006).

It is important to check for history of drug sensitivity and whether or not there is a family history consistent with unusual drug sensitivity or unusual reactions following anesthesia. Family history should be checked to determine the possibility of genetic conditions such as porphyria, or episodes of jaundice, possibly indicative of glucose-6-phosphate dehydrogenase deficiency.

Testing for Common Polymorphisms That Play a Role in Adverse Drug Reactions

Microarray analysis and specific mutation testing are being carried out more often, particularly in situations where patients may be exposed to a drug over a long period of time, for example, during treatment of tuberculosis, in cancer treatment, and in the case of long-term therapy with oral contraceptives and hormone replacement therapy. A single nucleotide polymorphism (SNP) set of probes for analysis of drug metabolism and transport was reported by Ahmadi et al. (2005). The microarray contains SNPs within 55 genes that are involved in drug metabolism, including genes encoding eight members of the cytochrome C family, 2 *N*-acetyltransferase genes, seven glutathione transferases, UDP glycosyl transferase, two paraoxonases, six dehydrogenases, four methyltransferases, sulfatase, and ATP-binding cassette proteins. In total, the microarray comprises 904 SNPs.

Polymorphisms Associated With Abnormal Sensitivity to Specific Pharmacological Agents

A few examples of common polymorphisms associated with abnormal response to a specific drug or class of drug are listed below.

Aminoglycoside sensitivity

A specific mutation in the mitochondrial genome in the *12SRNA* gene determines aminoglycoside sensitivity. The *A1555G* mutation is known to cause aminoglycoside hypersensitivity, also known as antibiotic-induced deafness. It is critically important to identify hypersensitive patients before they are placed at risk for hearing loss (Malik et al. 2003).

Patients with the *A1555G* mutation should avoid the use of aminoglycoside antibiotics. A single dose of aminoglycoside may be ototoxic to these patients; 100% of maternal family members are also at risk for sensorineural hearing loss. *A1555G* mutations can also cause nonsyndromic sensorineural hearing loss even in the absence of aminoglycoside exposure. In Spain, 15% to 20% of familial nonsyndromic hearing loss is due to this mutation. Many *A1555G*-positive patients can avoid hearing loss if protected from aminoglycoside exposure. Aminoglycoside antibiotics include: streptomycin, kanamycin, neomycin, gentamycin, tobramycin, apramycin, netilmycin, hygromycin B, spectinomycin, paromycin, and amikacin.

Hypercoagulable states and thrombotic risk due to variant factor V

A specific gene mutation in factor V, known as factor V Leiden, causes this coagulation factor to be inactivated at a rate approximately 10 times slower than normal factor V. The prolonged presence of the mutant factor results in increased thrombin generation. This in turn leads to a hypercoagulable state. Heterozygosity for the factor V mutation leads to a slightly increased risk for venous thrombosis; individuals who are homozygous for this mutation have a much greater thrombotic risk. The risk for thrombosis or thromboembolism is greatly increased in these individuals by hormone therapy, including contraceptive use and hormone replacement therapy.

Before prescribing these therapies, it is important to obtain a family history to determine if there are individuals who have had venous thrombosis. A coagulation screening test may provide some evidence for the disorder. Definitive diagnosis of this disorder may be made by DNA analysis (Franco and Reitsma 2001).

Venous thromboembolism is a well-known complication of oral contraception and hormonal replacement therapy. Inherited thrombophilia is viewed as an important determinant in modulating the effects of estrogens on thrombotic risk.

An increasing number of kits for thrombophilic mutations factor V Leiden (changes amino acid 506 from arginine to glutamine), prothrombin (*G20210A* gene mutation), and methylenetetrahydrofolate reductase (*C677T* gene change, or *A1298C*) are becoming commercially available, and screening for inherited thrombotic risk is among the most requested genetic tests in molecular diagnostic laboratories (Eldibany and Caprini 2007).

Genetic variation in drug sensitivity due to receptor polymorphisms

Response to inhaled corticosteroid therapy for asthma may differ significantly due to genetic variation in the cortisone receptor *CRHR1* (Tantisira et al. 2004).

Genetic variation in drug sensitivity due to ion channel polymorphisms

A common polymorphism, *S524Y*, in the *SCN5A* sodium channel gene impacts response to drugs used in the treatment of arrhythmias (Shuraih et al. 2007).

GENETIC VARIATION AND DRUGS USED IN CANCER TREATMENT

Thiopurine Methyltransferase Deficiency

In treatment of leukemias, 6-mercaptopurine or related drugs are sometimes used. Metabolism of this drug requires thiopurine methyltransferase (TPMT). TPMT is expressed in most tissues. It is polymorphic: 90% of individuals have high activity; 10% have intermediate activity; and 0.3% have very low or absent activity.

TPMT deficiency manifests as a codominant trait. Heterozygotes have lower levels of enzyme than normal and they have intermediate risks of hematological toxicity. TPMT-deficient patients have severe and sometimes fatal hematological sensitivity (Evans et al. 2001). At least 11 variant alleles lead to low TPMT activity. These alleles include single nucleotide changes leading to the following:

1. Amino acid substitutions that affect protein folding or protein function
2. Introduction of a stop codon and premature termination of transcription
3. Destruction of splice sites

Genetic Variants in Dihydropyrimidine Dehydrogenase

Fluorouracil treatment for malignancy can have severe toxic side effects, even at moderate doses, in patients who have mutations leading to reduced function of the enzyme dihydropyrimidine dehydrogenase. Heterozygotes and homozygotes are at risk for developing toxic side effects. Deficiency occurs in approximately 3% to 5% of the population and complete deficiency occurs in 0.2% (van Kuilenburg et al. 2003). Dihydropyrimidine dehydrogenase is the major enzyme responsible for reduction of fluorouracil to the inactive metabolite 5-6-dihydrofluorouracil. Many of the mutations leading to increased toxicity occur at sites in the enzyme that bind substrate or coenzyme. At least 30 variant alleles of dihydropyrimidine dehydrogenase occur. Morel et al. (2007) discovered a previously undescribed variant of this enzyme in a patient who died from polyvisceral toxicity in response to a first dose of fluorouracil.

PHARMACEUTICALS AND MOLECULAR-BASED CANCER THERAPY

In a review of the contribution of genetic studies to cancer therapy, Varmus (2006) expressed the opinion that for 50 years pharmaceutical chemistry served cancer patients more effectively than cancer biology; however, in the past decade genetic and biochemical studies have led to the development of less toxic forms of cancer drugs. These studies have revealed that changes in gene expression may cause tumors to become sensitive to specific hormone receptor antagonists or specific antibodies. The demonstration of altered cell signaling in cancer cells has led to development of specific pharmaceuticals that impact signaling pathways. He noted that despite successes with kinase inhibitors, relatively few drugs have been developed.

Varmus noted that there is evidence that many different types of cancer cells are oncogene dependent, so that the development of additional drugs that block oncogenes will be important. Another important consideration is that mutational hierarchies exist in cancers and successful treatment may not require that all mutations be targeted. Non-tumor-related therapies have a place in the treatment of cancer. Varmus emphasized antiangiogenesis therapies and antistromal cell therapies and consideration of the role of protease inhibitors (see Chapter 9 for further discussion of the molecular genetic basis of cancer therapy).

CLINICAL TRIALS

It is interesting to consider the standard description of protocols for clinical trials and the apparent absence of consideration as to how genetic variation in subjects will be assessed or taken into account in the evaluations of drug safety and drug efficacy.

Observational studies document signs and symptoms of disease, disease progression, and outcome (Gallin 2002).

In *interventional studies*, protocols are developed to measure the effect of a particular intervention on disease manifestations, progression, and outcome.

Clinical trials are designed to: (1) test safety, (2) compare the efficacy of a new therapy to that of an old therapy, and (3) test efficacy of the new therapy. Clinical trials involve not only the use of medications but also testing of devices, procedures, and other therapeutic interventions. Clinical trials may also be designed to test the efficacy of a method to screen for a disease. Trials may be designed to test methods of disease prevention. In the case of genetic diseases, this could involve methods to test for the presence of a disease-causing mutant gene prior to the occurrence of symptoms of the disease.

Diagnostic trials involve investigation of better methodologies that lead to more accurate diagnosis of specific diseases. The best methods of supportive care may also be investigated.

Therapeutic trials include randomized control trials, double-blind trials, or active control trials where one group is given the test drug and another is given a previously investigated drug active against the specific disease.

A phase 1 trial is a safety trial carried out on healthy volunteers, to best define dosage, timing, and mode of administration. Phase 2 involves studies on 20–300 individuals to test treatment efficacy. Phase 3 involves studies on a larger patient group than that studied in phase 2. Phase 4 is a postlaunch safety trial. Prior to initiating a trial, the institutional review board reviews protocols and ethics.

Questions relating to the subjects' welfare must be asked throughout the trial. Data monitoring committees must be established for ongoing examination of safety and record keeping, quality of data collected, accuracy and timeliness of data collection, and transparency of the trial conduct.

Power refers to the ability to detect a statistically significant difference between treated and control groups or difference within the group receiving the new treatment versus the standardized treatment group.

DRUG DESIGN: TARGETING DOWNSTREAM EFFECTS OF GENE MUTATIONS AND USE OF TRANSGENIC MOUSE MODELS OF HUMAN GENETIC DISEASES

Through characterization of gene defects leading to specific phenotypic abnormalities, analysis of gene function, and elucidation of the downstream effects of specific gene mutations, we gain insight into steps that can be impacted by therapeutic agents. Examples of application of analysis of molecular pathogenesis to development of therapies are presented below for Marfan syndrome (MFS) and achondroplasia.

Marfan Syndrome: Targeting Downstream Effects of Mutation

Marfan syndrome is an autosomal dominant disorder characterized by increased growth of long bones, reduced muscle mass, ocular lens dislocation, dissecting aneurysm of the aorta, and cardiac disease, often associated with mitral valve anomalies and arrhythmias. In 25% of cases there is no family history and the condition is de novo. This disorder is due to mutations in the fibrillin-encoding gene (*FBN1*) on chromosome 15q21.1 (Dietz and Pyeritz 1995). Neptune et al. (2003) determined that fibrillin is a negative regulator of transforming growth factor beta (TGF-beta). The discovery that mutations in TGF-beta receptors 1 and 2 on chromosome 3 lead to vascular abnormalities similar to those found in Marfan syndrome provides additional evidence for the role of TGF-beta in inducing Marfan syndrome and Marfan syndrome–like cardiovascular anomalies.

The mutations in the *FBN1* gene in Marfan syndrome impair the ability of fibrillin to negatively regulate TGF-beta activity. Neptune et al. (2003) and Ng et al. (2004) demonstrated that TGF-beta overactivity is a key in determining aortic aneurysm formation, aortic root dilation, and valve disease in Marfan syndrome.

Discovery of the downstream effects of fibrillin mutations and the specific pathogenesis of aorta, mitral valve, and muscle disease in Marfan syndrome was made possible through development of a transgenic mouse model of the disease. Judge et al. (2004) reported that the characteristic features of Marfan syndrome are present in mice heterozygous for a *C1039G* mutation. Studies on these mice revealed that anti-TGF-neutralizing antibody prevented development of severe pathology.

Antihypertensive drugs are frequently used in the treatment of Marfan syndrome to reduce the risk of aortic rupture. Drugs used include beta-blockers and angiotensin-converting enzyme inhibitors. Habashi et al. (2006) reported that the angiotensin II type 1 receptor blocker losartan prevented development of aortic root dilation and aneurysm in the mouse model of Marfan syndrome.

Marfan syndrome is characterized by muscle hypoplasia and hypotonia. Studies by Cohn et al. (2007) in the Marfan mouse model revealed that abnormal muscle regeneration is a key problem. This is due to a decline in the proliferation and function of satellite cells. Cohn et al. demonstrated that TGF-beta-neutralizing antibody and losartan treatment enhanced muscle regeneration. They also noted that treatment with these agents enhanced satellite cell performance and muscle regeneration in the Marfan mouse model.

Achondroplasia: Need for Targeted Delivery of Therapeutic Agent

In many disorders due to gene mutation, the aberrant function of the mutant protein particularly impacts one tissue or organ. A problem that emerges then in designing therapy is that the therapeutic agent needs to be delivered primarily to tissue type or organ. In cancer therapy there are examples of development of antibodies and drug-coupled antibodies to specifically target a tumor-specific protein. A goal for the future will be development of tissue-targeted reagents for specific genetic disorders. One possibility under consideration is the development of drugs targeted to cartilage and bone growth plates to decrease aberrant activation of the *FGFR3* receptor in achondroplasia (Horton et al. 2007).

Achondroplasia is characterized by short-limbed dwarfism and vertebral anomalies that may lead to spinal stenosis. It is due to mutation in the gene that encodes fibroblast growth factor receptor, *FGFR3*. The same mutation occurs in 95% of patients with this disorder. De novo mutations occur in 80% of cases.

Normal signaling through *FGFR3* results in inhibition of proliferation and terminal differentiation of growth plate chondrocytes. Exaggerated signaling through the mutant *FGFR3* occurs in achondroplasia and results in impaired bone growth. Small chemical inhibitors of *FGFR3* tyrosine kinase activity have been developed. However, although they are effective in tissue culture, they are not effective in animal studies. Horton et al. (2007) noted that highly specific humanized antibodies to *FGFR3* have been developed. It is not clear to what extent these will target chondrocytes.

FGF binding to *FGFR3* leads to dimerization of the receptor monomers and activation of the receptor kinase function. This in turn activates a number of downstream signaling pathways including the *MAPK* (mitogen activated kinase) pathway and the *STAT1* (signal transducer and activator of transcription) pathway (Dailey et al. 2003). The *MAPK* pathway is inhibited by the C-type natriuretic peptide CNP, which binds to the natriuretic peptide receptor B (NPRB). CNP and NPRB are expressed in the bone growth plate. Possibilities of treatment of achondroplasia with CNP or CNP analogues are being evaluated (Horton et al. 2007).

Organelle Targeting

Ubiquinones protect mitochondria from oxidative damage, but their uptake in mitochondria is minimal. Cocheme et al. (2007) developed a process in which ubiquinone is conjugated to lipophilic triphenylphosphonium cation to produce a compound known as MitoQ. This compound passes through membranes and accumulates at high concentration in mitochondria. There it protects against oxidative damage.

11

PHENOMES, PHENOTYPES, AND CLINICAL GENETIC SERVICES

> As genomic research moves toward proteomics and proteomics moves toward proteotype phenotype correlations, clarity concerning clinical features will become even more important.
>
> —Judith Hall

Gene function may be speculated on based on sequence, domain structure, and homology to other sequences. Hall (2003) emphasized that clues about the function of a gene may be forthcoming from clear clinical descriptions of patients who are found to have mutations. She noted in addition that descriptions on longitudinal studies in such patients are important because physical features change over time and different problems emerge as these patients age. Hall has drawn attention to the fact that lack of detailed clinical information will compromise progress in the understanding of biology and disease.

Feenstra et al. (2006) emphasized that description of phenotypic features is subjective. A number of proposals have been put forth to standardize clinical descriptions of human malformations. Photographs are valuable. However, development of computer programs to analyze craniofacial features based on three-dimensional imaging leads to greater accuracy of syndrome identification (Loos et al. 2003).

A PHENOME PROJECT

In 2003, Freimer and Sabatti proposed that an international effort be under-taken to create phenotype or phenomic databases and to develop new ap-proaches for analyzing phenotype data. They emphasized that progress in identifying the genetic basis of common diseases is delayed because of the manner in which the phenotype is described. They noted that phenotype de-scriptions rooted in medical traditions are variable. They called for stan-dardized comprehensive databases with phenotypic descriptions.

The phenotype includes morphological, biochemical, physiological, and behavioral characteristics of an organism. Freimer and Sabatti noted further that phenome characterization includes information at different levels of res-olution, of the organism, tissues, cells, and molecules. Another form of in-formation to be included relates to exposures.

In reports of genetic diseases, information on the phenotype is often contained within the text and is not readily organized for computer analysis. A number of investigators have developed text mining methods (Masseroli et al. 2005; Masseroli and Pinciroli 2006).

BIOBANKS

Biobank-related efforts aim to integrate molecular information from biolog-ical samples with clinical phenotype information and information on lifestyle and environmental exposures. One such database originated in the Avon project (Pembrey 2004).

Muilu et al. (2007) described a federated database for integrating phe-nome and genome data from 600,000 twins. They defined a federated database as a data management system that integrates multiple autonomous databases into a single database.

IMPORTANCE OF OBSERVATION AND DOCUMENTATION FOR PHENOTYPE ANALYSIS

Technological developments allow more comprehensive analysis of the phe-notype in some cases. Nevertheless, observational skills at the time of initial contact and elicitation of family history as well as ongoing observation and careful documentation are critical for diagnosis and management. In some cases, the phenotype and progression may not fit into a known category.

However, careful observation and documentation of phenotype along with genome and gene analysis and consideration of related entities will be critical in the elucidation of new entities.

Accurate phenotypic description requires clinical training and application of skills. Documentation of the unusual requires time and patience. The question arises as to whether economic and cultural conditions within medical practice are conducive to or even permit the type of dedication required for careful description and documentation of phenotype. In addition, it has become more difficult to publish journal articles that include detailed patient descriptions.

IDENTIFYING DISEASE GENES AND CORRELATING GENES AND PHENOTYPE

Van Driel et al. (2006) drew attention to the value of grouping disease phenotypes into a matrix to predict biological relations between genes and proteins. They noted that individual genes that cause a specific phenotype are often linked at a biological level.

There are many examples of disorders where functionally related genes induce similar phenotypes. Oti et al. (2007) reviewed several of these. They noted that the biological relationship of genes might be based on the fact that they function in a multiprotein complex or because they function in the same pathway. In some cases, different genes that function within the same organelle give rise to similar symptoms. Based on these premises, bioinformatics methods may be useful in identifying related genes. Specific databases may be used to identify functionally related genes. Yeast two hybrid systems are valuable for characterization of interacting proteins.

Mutations in Genes That Encode Different Members of a Protein Complex Lead to Similar Phenotypes

Fanconi anemia

Fanconi anemia is one example of a disorder in which interacting genes give rise to a specific phenotype that includes progressive aplastic anemia, multiple congenital anomalies, especially limb abnormalities, and predisposition to leukemia and solid tumors. In response to irradiation or UV exposure–induced damage of DNA, genes in this specific pathway coordinate DNA repair, recombination, and replication (Mathew 2006). Seven different

Fanconi anemia genes are assembled in a nuclear core complex. This core complex then activates Fanconi D2 and this gene then initiates DNA replication.

Bardet-Biedl syndrome

Bardet-Biedl syndrome (*BBS1*) is caused by mutations in any one of 12 different genes. The syndrome is inherited as an autosomal dominant. Clinical features of this syndrome include cone-rod dystrophy of the retina that may lead to blindness, cognitive impairment, postaxial polydactyly, truncal obesity, hypogonadism, genitourinary tract anomalies, and renal anomalies. Mutations in the *BBS1* gene occur in 18% to 32% of patients; *BBS10* mutations occur in approximately 10% of patients.

Nachury et al. (2007) reported that dysfunction of primary cilia underlies this syndrome. Seven BBS proteins form a core structure referred to as the BBSome, which localizes to cilia membranes and is required for ciliogenesis. The BBSome is in contact with Rab8 GDP GTP exchange factor, which plays a key role in extension of the ciliary membrane. Nachury et al. proposed that the BBSome together with the Rab8 GDP GTP exchange factor plays a role in vesicular trafficking (Figure 11–1).

Mutations in Genes That Encode Functionally Related Proteins Lead to a Similar Phenotype

Cornelia de Lange syndrome

Cornelia de Lange syndrome (CDLS) is a multisystem developmental disorder that leads to malformation and to distinct facial features that include low anterior hairline, prominent arched brows that meet in the middle (synophrys), upturned nose with depressed nasal bridge, prominent upper jaw, thin lips, and long philtrum. Malformations frequently involve the upper extremities, particularly the hands, and gastroesophageal abnormalities occur. Growth retardation and various degrees of mental retardation are common in these patients. Variant, often milder, forms of CDLS have been described by a number of investigators.

Tonkin, Smith, et al. (2004) and Tonkin, Wang, et al. (2004) determined that one copy of the gene *NIPBL* (the human homologue of *Drosophila* nipped gene) on chromosome 5p13.1 was disrupted or mutated in patients with CDLS. Gillis et al. (2004) reported that *NIPBL* mutations occur in approximately 50% of cases of CDLS. *NIPBL* is the homologue of the yeast gene *Sec2* that regulates the loading of cohesins onto chromosomes. Cohesin plays a role in chromatid cohesion and coordinated segregation of chomatids during cell

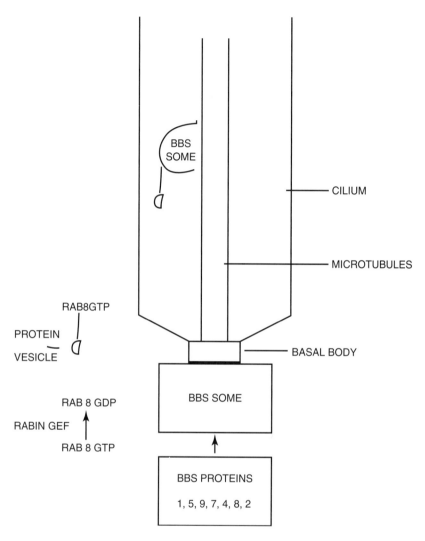

Figure 11–1. Mutations in a number of different genes can lead to Bardet-Biedl syndrome. At least seven of these genes encode proteins that form a complex known as the BBS some (BBSome). This structure promotes cilia formation and molecular movement along the microtubular structures within cilia (based on Nachury et al. 2007).

division. Cohesin also plays a role in the repair of DNA breaks. Cohesin binding takes place at different positions along the chromosome. Dorsett et al. (2005) reported evidence that cohesin binding regulates gene expression. They demonstrated that binding of cohesin at a position between a specific promoter and enhancer downregulates expression of the *Drosophila* cut gene.

Cohesin is a multiprotein complex that contains subunits of structural maintenance chromosome proteins, encoded by the genes *SMC1A* and *SMC3*. These are present at the core of the cohesin complex. This complex also contains the protein encoded by *SCC1* (also known as *RAD21*) and *SA1* and *SA2* (stromal antigen 1 and 2, sometimes known as *STAG1* and *STAG2*). The gene that encodes *SMC1A* maps to the X chromosome and the *SMC3* gene maps to chromosome 10q24-25 (Deardorff et al. 2007).

Mutation or disruption of the *SMC1A* locus on chromosome Xp11.22 leads to CDLS manifestations. Musio et al. (2006) reported that CDLS occurred in a female patient where an apparently balanced translocation disrupted the *SMC1A* gene. They also identified *SMC1A* mutations in four male patients with CDLS. One male was apparently a sporadic case; three males were members of a single family. Deardorff et al. (2007) described mutations in the *SMC3* encoding gene, which maps on chromosome 10q24-q25, in patients with CDLS. The phenotype in these patients was milder than that seen in the patients with *NIPBL* mutations. Specifically, major abnormalities in limbs and gastrointestinal tract were less common, and growth and mental retardation were less severe.

Roberts syndrome and SC phocomelia syndrome

The ESCO2 protein (establishment of cohesin) plays a role in cohesin loading and chromatid cohesion. This protein is mutated in some cases of Roberts syndrome and the SC phocomelia syndrome, also known as the pseudothalidomide syndrome (Vega et al. 2005). Roberts syndrome shows phenotypic variability. The manifestations include limb abnormalities, hypoplasia or aplasia of the radius or fibula, and clubbed foot. Chromosome abnormalities include loss of cohesion of centromeres at the heterochromatic region. The SC phocomelia syndrome is characterized by limb reduction abnormalities, growth retardation, and in some instances mental retardation. Chromosome abnormalities include premature centromere separation and chromatid repulsion. The ESCO2 protein also has acetyltransferase activity and is required for chromatid cohesion.

These findings indicate how mutations in different members of a protein complex, such as *SMC1A* and *SMC3*, and mutations in a protein responsible for localizing that complex to its target lead to syndromes with similar features. Furthermore, these reports indicate how discovery of the genes responsible for genetic syndromes and analyses of the proteins they encode provide new insights into physiological processes.

IMPORTANCE OF ONGOING CLINICAL RESEARCH FOLLOWING DISEASE GENE IDENTIFICATION

Neurofibromatosis: Ongoing Research and Translational Relevance

Neurofibromatosis is characterized by the development of tumors, primarily neurofibromas that arise in the skin and in the peripheral and central nervous system. Pigment changes commonly occur in this disorder, café au lait spots and axillary freckling. Eye changes (Lisch nodules) and bony changes (erosions) are often present. Vascular changes may lead to strokes and cerebral ischemia. Neurofibromatosis (*NF1*) occurs with a frequency of 1 in 3000. It is inherited as an autosomal dominant condition; however, 30% to 50% of cases arise as new mutations.

The *NF1* gene maps to chromosome 17q11.2 and encodes neurofibromin. Disease-causing mutations arise throughout the gene. They may be single nucleotide mutations or deletions. Identification of the causative mutation in a specific *NF1* patient is hampered by the large size of the gene, the fact that mutations may arise throughout the gene, and the presence of many pseudogenes. It is interesting to note that children with features of *NF1* and with hematologic disorders have been found to have mutations in the mismatch repair gene *MLH1* (Wang et al. 2003) and *MSH6* (Hegde et al. 2005).

Complex Syndromes With Features of Neurofibromatosis

NF1 deletion syndrome

Large deletions involving the *NF1* gene and flanking genes give rise to the *NF1* deletion syndrome. This syndrome is characterized by features of *NF1* and unusual facial features (dysmorphology), joint laxity, and delayed development. The causative deletions often have breakpoints in low copy repeats in the *NF1* flanking regions, and at least 14 genes may be deleted in these type 1 deletions (Mensink et al. 2006). In some cases, deletion is caused by abnormal recombination between the *JJAZ1* (Zeste suppressor) gene and its pseudogene. These deletions are referred to as type 2 deletions. Type 1 deletions may arise during interchromosomal recombination during meiosis. Type 2 deletions may arise as a result of intrachromosomal recombination during mitosis.

NF1-Noonan syndrome

A subgroup of patients with *NF1* have congenital anomalies similar to those found in Noonan syndrome (Kehrer-Sawatzki et al. 2004). These include congenital heart anomalies, pulmonary stenosis, aortic insufficiency, and unusual facial features. Classic Noonan syndrome is an autosomal dominant condition characterized by minor facial anomalies including hypertelorism, downward slant of palpebral fissures, ptosis, sternal abnormalities (pectus excavatum), webbed neck, learning disabilities, and congenital heart anomalies. The heart anomalies include atrioventricular canal defects, pulmonary artery stenosis, ventricular septal defect, and atrial septal defect. Noonan syndrome is genetically heterogeneous and 40% of cases are due to mutation in the *PTPN11* gene on chromosome 12q24.1. This syndrome may arise due to mutation in the *KRAS* gene on chromosome 12p12.1.

Huffmeier et al. (2006) postulated that constitutive deregulation of the *RAS* pathway leads to Noonan syndrome. In cases that had *NF1* mutations and Noonan syndrome, there was clustering of *NF1* mutations in the RAS-GTPase domain of *NF1*, between exons 21 and 27. *PTPN11* is an upstream activator of the *RAS* pathway. *PTPN11* encodes a protein tyrosine phosphatase (*PTP*) and two *SH2* domains. This protein, referred to as SHP2, specifically phosphorylates tyrosine residues on cell surface receptors and cell adhesion molecules. Through this phosphotyrosine modulation, it enhances intracellular signal transduction through the *RAS* and *MAPK* pathways.

GENETIC SERVICES, PERSONNEL, AND TRAINING

Types of Service

The question arises as to who should provide services for patients with genetic diseases. Donnai (2002) reviewed genetic services in the United Kingdom and concluded that genetic services should be considered in two separate categories. The first involves services that target whole populations in order to identify individuals at risk. Newborn screening represents one example of this. A second category of services focuses on the need of families affected by a genetic disorder or at high risk for disorder.

For the second category, the question arises whether the genetics specialist plays a role in the initial diagnosis only or whether the specialist is involved at some level, working with a primary care physician in lifelong coordinated care. For care of patients with inborn errors of metabolism, specialist care centers with multidisciplinary teams are more likely to be involved.

Personnel

It is important that primary care physicians and nurse practitioners become increasingly skilled in identification of genetic disorders and of risks within families. Donnai (2002) noted that genetics will not remain the prerogative of regional or academic genetic centers. All branches of medicine will need to use genetic knowledge in their practice.

Genetic screening and counseling are becoming increasingly important in care of cancer patients, particularly in cases where there is a family history of cancer. Henriksson et al. (2004) reported experiences in oncogenetic counseling in a clinic in Lund, Sweden, over a 10-year period. They identified 789 families with a strong family history of cancer. In 76 families there was a family history of breast and ovarian cancer. In 45 of these families a cancer-predisposing mutation was identified. Henriksson et al. encountered 129 families with a strong family history of breast cancer; cancer-predisposing mutations were identified in 27 of these. Microsatellite instability occurred in tumors of 34 patients with a family history of colon cancer, and in 16 of these families a germline cancer-predisposing mutation was identified. These results indicate the importance of identifying families where increased surveillance is necessary to identify malignancies at an early stage.

Donnai (2002) considered the role of genetic registries, which in the UK health care setting serves to promote ongoing contact with families and also contact with extended families.

Training

In most medical school curricula, genetics is taught as a single course. It is important to consider that genetics should be part of the training throughout the curriculum. Genetics should also be a component of training in all medical specialties and should be included in continuing education.

Increasingly, clinical scientists express concerns about training and integration of physicians in research. Fears et al. (1999) promoted the concept of interdisciplinary frameworks to integrate molecular biology and clinical genetics.

MULTICULTURAL AND MULTIETHNIC SOCIETIES
AND ACCESS TO GENETIC SERVICES

In cities and regions with large multiethnic, multicultural societies, there are special considerations in providing genetic services. Population differences in the prevalence of specific genetic disorders, for example, hemoglobinopathies, need to be taken into account (Weatherall 2005). There are population differences with respect to pharmacogenetic responses that require attention.

Significant differences in religious, cultural, and societal structures exist and require consideration. Mehta and Saggar (2005) emphasized that in planning genetic services in such communities, language diversity and values diversity require consideration. Furthermore, they recommended wider community, public, and patient engagement in the provision of genetic services and the inclusion of diverse ethnic populations in health research.

Modell and Darr (2002) and Weatherall (2005) noted that the genetic implications of consanguineous marriages and recessive genetic diseases have come to attention in the Middle East and South Asia as infant mortality due to infectious diseases and nutritional deficiency has declined. In addition, these problems are coming more to the foreground in Europe as the immigrant populations rise.

ESTABLISHING EVIDENCE-BASED GUIDELINES
AND PROTOCOLS FOR MANAGEMENT

Campbell et al. (2000) described processes to develop evidence-based national clinical guidelines and integrated care pathways for five genetic conditions that represent 30% of all clinical genetic consultations in Scotland. These conditions included tuberous sclerosis, myotonic dystrophy, Marfan syndrome, Huntington disease, and neurofibromatosis type 1.They also reviewed evidence that supported the proposed clinical management recommendations. The integrated care pathway format used for patient records provided a structure to capture the key clinical data and served as a reminder of recommendations for clinical management. Campbell et al. proposed ongoing audits and continued research on the development of the best clinical guidelines for diagnosis and management of patients with genetic conditions.

REVIEW OF GENETIC SERVICES IN EUROPE

Godard et al. (2003) undertook a comprehensive review of genetic services throughout Europe. This report led to an appeal for the establishment of best practice guidelines, for harmonization of rules within Europe for coverage for genetic tests. The authors of the report promoted the concept of regionalization for provision of care. The group raised concerns about inequality of access to genetic services and differences in the effectiveness of services. They stressed the importance of multidisciplinary services for patients with genetic disorders.

Included in categories of genetic services were diagnostic consultations, information and counseling, prenatal diagnosis, cascade carrier testing (testing the family of an affected individual), and presymptomatic and predictive testing. This review emphasized the psychological complexity of presymptomatic and predictive testing and that this required a multidisciplinary approach.

The reviewers recommended use of systems of clinical audit to monitor protocols of care, quality of record keeping, and quality and promptness of explanatory letters to referring physicians and patients. They recommended that a system to monitor follow-up with patients regarding test results be in place and that units should have systems in place to monitor effectiveness of genetic services from the points of view of both patients and professionals.

The report from Europe opined, "There is a consensus in order that the costs of specialist genetic services be collectively covered by the public health system, health insurance or other means used in the country concerned. The costs should not be left to the individual family and it may be unfair to leave them to a local small community" (Godard et al. 2003, p. S23).

The group considered the movement of research to the clinic. They concluded, "Mechanisms need to be developed to translate beneficial research findings in a framework which allows for evaluation and further development" (p. S22).

The Godard et al. report emphasized that the key difficulty that arises in presymptomatic testing is that physicians often have no recommendations for cure or for prevention of disease symptoms.

CONCLUSION

Based on the material presented in the preceding chapters, one must conclude that there are encouraging developments in the area of therapeutics in genetics.

MANAGEMENT OF GENETIC DISEASES

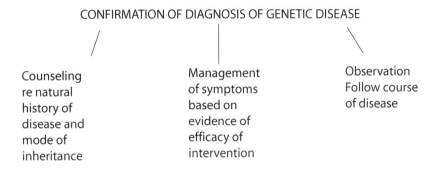

CONFIRMATION OF DIAGNOSIS OF GENETIC DISEASE

Counseling
re natural
history of
disease and
mode of
inheritance

Management
of symptoms
based on
evidence of
efficacy of
intervention

Observation
Follow course
of disease

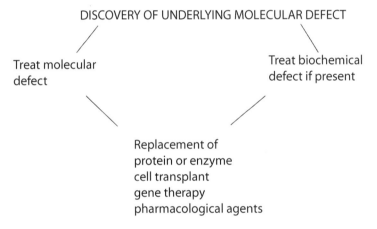

DISCOVERY OF UNDERLYING MOLECULAR DEFECT

Treat molecular
defect

Treat biochemical
defect if present

Replacement of
protein or enzyme
cell transplant
gene therapy
pharmacological agents

Figure 11–2. Approaches to the management and treatment of genetic disease are impacted by discovery of the underlying molecular defect and molecular pathogenesis.

Until recently, much of genetic medicine was concentrated on providing individuals and families with information concerning genetic risk and on dealing with treatment of symptoms. Advances in molecular biology, availability of the human genome sequences, and the capacity to model genetic diseases in transgenic mice, for example, have led to an explosion of information on underlying disease mechanisms and pathogenesis for a large and growing number of genetic diseases. Elucidation of molecular mechanisms has led to development of specific therapeutic options for a number of diseases (Figure 11–2). There is promise for continued development of therapeutics and progress in delineating steps for prevention of symptoms in patients with a genetic predisposition to development of late-onset disorders.

REFERENCES

Aardema MJ, MacGregor JT. 2002. Toxicology and genetic toxicology in the new era of "toxicogenomics": Impact of "-omics" technologies. *Mutat Res* 499(1):13–25.

Abadie V, Berthelot J, Feillet F, Maurin N, Mercier A, de Baulny HO, de Parscau L. 2001. Neonatal screening and long-term follow-up of phenylketonuria: The French database. *Early Hum Dev* 65(2):149–58.

Adler EM, Gough NR. 2007. Focus issue: Exploring new avenues for cancer treatment. *Sci STKE* 2007(381):eg2.

Aharon-Peretz J, Rosenbaum H, Gershoni-Baruch R. 2004. Mutations in the glucocerebrosidase gene and Parkinson's disease in Ashkenazi Jews. *N Engl J Med* 351(19):1972–7.

Ahmadi KR, Weale ME, Xue ZY, Soranzo N, Yarnall DP, Briley JD, Maruyama Y, Kobayashi M, Wood NW, Spurr NK, and others. 2005. A single-nucleotide polymorphism tagging set for human drug metabolism and transport. *Nat Genet* 37(1): 84–9.

Aitman TJ, Dong R, Vyse TJ, Norsworthy PJ, Johnson MD, Smith J, Mangion J, Roberton-Lowe C, Marshall AJ, Petretto E, and others. 2006. Copy number polymorphism in Fcgr3 predisposes to glomerulonephritis in rats and humans. *Nature* 439(7078):851–5.

Akiva P, Toporik A, Edelheit S, Peretz Y, Diber A, Shemesh R, Novik A, Sorek R. 2006. Transcription-mediated gene fusion in the human genome. *Genome Res* 16(1): 30–6.

Allen KM, Gleeson JG, Bagrodia S, Partington MW, MacMillan JC, Cerione RA, Mulley JC, Walsh CA. 1998. PAK3 mutation in nonsyndromic X-linked mental retardation. *Nat Genet 20*(1):25–30.

Anderson SL, Coli R, Daly IW, Kichula EA, Rork MJ, Volpi SA, Ekstein J, Rubin BY. 2001. Familial dysautonomia is caused by mutations of the IKAP gene. *Am J Hum Genet 68*(3):753–8.

Bailey JA, Gu Z, Clark RA, Reinert K, Samonte RV, Schwartz S, Adams MD, Myers EW, Li PW, Eichler EE. 2002. Recent segmental duplications in the human genome. *Science 297*(5583):1003–7.

Balkwill F, Mantovani A. 2001. Inflammation and cancer: Back to Virchow? *Lancet 357*(9255):539–45.

Barnetson RA, Tenesa A, Farrington SM, Nicholl ID, Cetnarskyj R, Porteous ME, Campbell H, Dunlop MG. 2006. Identification and survival of carriers of mutations in DNA mismatch-repair genes in colon cancer. *N Engl J Med 354*(26): 2751–63.

Bartolomei MS, Tilghman SM. 1997. Genomic imprinting in mammals. *Annu Rev Genet 31*:493–525.

Bayani J, Selvarajah S, Maire G, Vukovic B, Al-Romaih K, Zielenska M, Squire JA. 2007. Genomic mechanisms and measurement of structural and numerical instability in cancer cells. *Semin Cancer Biol 17*(1):5–18.

Bear MF, Huber KM, Warren ST. 2004. The mGluR theory of fragile X mental retardation. *Trends Neurosci 27*(7):370–7.

Beck M. 2007. New therapeutic options for lysosomal storage disorders: Enzyme replacement, small molecules and gene therapy. *Hum Genet 121*(1):1–22.

Begue B, Dumant C, Bambou JC, Beaulieu JF, Chamaillard M, Hugot JP, Goulet O, Schmitz J, Philpott DJ, Cerf-Bensussan N, and others. 2006. Microbial induction of CARD15 expression in intestinal epithelial cells via toll-like receptor 5 triggers an antibacterial response loop. *J Cell Physiol 209*(2):241–52.

Bell RD, Sagare AP, Friedman AE, Bedi GS, Holtzman DM, Deane R, Ziokovic BV. 2007. Transport pathways for clearance of human Alzheimer's amyloid beta-peptide and apolipoproteins E and J in the mouse central nervous system. *J Cereb Blood Flow Metab 27*(5) 909–18.

Bender A, Krishnan KJ, Morris CM, Taylor GA, Reeve AK, Perry RH, Jaros E, Hersheson JS, Betts J, Klopstock T, and others. 2006. High levels of mitochondrial DNA deletions in substantia nigra neurons in aging and Parkinson disease. *Nat Genet 38*(5):515–7.

Bennett ST, Lucassen AM, Gough SC, Powell EE, Undlien DE, Pritchard LE, Merriman ME, Kawaguchi Y, Dronsfield MJ, Pociot F, and others. 1995. Susceptibility to human type 1 diabetes at IDDM2 is determined by tandem repeat variation at the insulin gene minisatellite locus. *Nat Genet 9*(3):284–92.

Benvenuto G, Li S, Brown SJ, Braverman R, Vass WC, Cheadle JP, Halley DJ, Sampson JR, Wienecke R, DeClue JE. 2000. The tuberous sclerosis-1 (TSC1) gene product hamartin suppresses cell growth and augments the expression of the TSC2 product tuberin by inhibiting its ubiquitination. *Oncogene 19*(54):6306–16.

Bergmann C, Frank V, Kupper F, Schmidt C, Senderek J, Zerres K. 2006. Functional analysis of PKHD1 splicing in autosomal recessive polycystic kidney disease. *J Hum Genet 51*(9):788–93.

Blencowe BJ. 2006. Alternative splicing: New insights from global analyses. *Cell* *126*(1):37–47.

Bommer UA, Thiele BJ. 2004. The translationally controlled tumour protein (TCTP). *Int J Biochem Cell Biol 36*(3):379–85.

Bondi ML, Craparo EF, Giammona G, Cervello M, Azzolina A, Diana P, Martorana A, Cirrincione G. 2007. Nanostructured lipid carriers containing anticancer compounds: Preparation, characterization, and cytotoxicity studies. *Drug Deliv 14*(2): 61–7.

Bordet R, Ouk T, Petrault O, Gele P, Gautier S, Laprais M, Deplanque D, Duriez P, Staels B, Fruchart JC, and others. 2006. PPAR: A new pharmacological target for neuroprotection in stroke and neurodegenerative diseases. *Biochem Soc Trans 34*(Pt 6):1341–6.

Botstein D, Risch N. 2003. Discovering genotypes underlying human phenotypes: Past successes for mendelian disease, future approaches for complex disease. *Nat Genet 33* Suppl:228–37.

Brandwijk RJ, Griffioen AW, Thijssen VL. 2007. Targeted gene-delivery strategies for angiostatic cancer treatment. *Trends Mol Med 13*(5):200–9.

Brichta L, Hofmann Y, Hahnen E, Siebzehnrubl FA, Raschke H, Blumcke I, Eyupoglu IY, Wirth B. 2003. Valproic acid increases the SMN2 protein level: A well-known drug as a potential therapy for spinal muscular atrophy. *Hum Mol Genet 12*(19): 2481–9.

Brody LC. 2005. Treating cancer by targeting a weakness. *N Engl J Med 353*(9): 949–50.

Brooks DA. 2007. Getting into the fold. *Nat Chem Biol 3*(2):84–5.

Buolamwini JK, Addo J, Kamath S, Patil S, Mason D, Ores M. 2005. Small molecule antagonists of the MDM2 oncoprotein as anticancer agents. *Curr Cancer Drug Targets 5*(1):57–68.

Bykov VJ, Issaeva N, Selivanova G, Wiman KG. 2002. Mutant p53-dependent growth suppression distinguishes PRIMA-1 from known anticancer drugs: a statistical analysis of information in the National Cancer Institute database. *Carcinogenesis 23*(12):2011–8.

Bykov VJ, Issaeva N, Zache N, Shilov A, Hultcrantz M, Bergman J, Selivanova G, Wiman KG. 2005. Reactivation of mutant p53 and induction of apoptosis in human tumor cells by maleimide analogs. *J Biol Chem 280*(34):30384–91.

Calabrese C, Poppleton H, Kocak M, Hogg TL, Fuller C, Hamner B, Oh EY, Gaber MW, Finklestein D, Allen M, and others. 2007. A perivascular niche for brain tumor stem cells. *Cancer Cell 11*(1):69–82.

Campbell H, Bradshaw N, Davidson R, Dean J, Goudie D, Holloway S, Porteous M. 2000. Evidence based medicine in practice: Lessons from a Scottish clinical genetics project. *J Med Genet 37*(9):684–91.

Canevari L, Clark JB. 2007. Alzheimer's disease and cholesterol: The fat connection. *Neurochem Res 32*(4–5):739–50.

Cao H, Jung M, Stamatoyannopoulos G. 2005. Hydroxamide derivatives of short-chain fatty acid have erythropoietic activity and induce gamma gene expression in vivo. *Exp Hematol 33*(12):1443–9.

Caparros-Lefebvre D, Lees AJ. 2005. Atypical unclassifiable parkinsonism on Guadeloupe: An environmental toxic hypothesis. *Mov Disord 20* Suppl 12:S114–8.

Cardon LR. 2006. Genetics. Delivering new disease genes. *Science 314*(5804): 1403–5.

Carninci P. 2006. Tagging mammalian transcription complexity. *Trends Genet 22*(9): 501–10.

Caron PR, Mullican MD, Mashal RD, Wilson KP, Su MS, Murcko MA. 2001. Chemogenomic approaches to drug discovery. *Curr Opin Chem Biol 5*(4):464–70.

Carson, R. 1962. *Silent Spring.* New York: Houghton Mifflin.

Cartegni L, Krainer AR. 2003. Correction of disease-associated exon skipping by synthetic exon-specific activators. *Nat Struct Biol 10*(2):120–5.

Castellone MD, Teramoto H, Williams BO, Druey KM, Gutkind JS. 2005. Prostaglandin E2 promotes colon cancer cell growth through a Gs-axin-beta-catenin signaling axis. *Science 310*(5753):1504–10.

Castle J, Garrett-Engele P, Armour CD, Duenwald SJ, Loerch PM, Meyer MR, Schadt EE, Stoughton R, Parrish ML, Shoemaker DD, and others. 2003. Optimization of oligonucleotide arrays and RNA amplification protocols for analysis of transcript structure and alternative splicing. *Genome Biol 4*(10):R66.

Chalmers K, Wilcock GK, Love S. 2003. APOE epsilon 4 influences the pathological phenotype of Alzheimer's disease by favouring cerebrovascular over parenchymal accumulation of A beta protein. *Neuropathol Appl Neurobiol 29*(3):231–8.

Chan AT, Ogino S, Fuchs CS. 2007. Aspirin and the risk of colorectal cancer in relation to the expression of COX-2. *N Engl J Med 356*(21):2131–42.

Chen X, Truong TT, Weaver J, Bove BA, Cattie K, Armstrong BA, Daly MB, Godwin AK. 2006. Intronic alterations in BRCA1 and BRCA2: Effect on mRNA splicing fidelity and expression. *Hum Mutat 27*(5):427–35.

Cheng J, Kapranov P, Drenkow J, Dike S, Brubaker S, Patel S, Long J, Stern D, Tammana H, Helt G, and others. 2005. Transcriptional maps of 10 human chromosomes at 5-nucleotide resolution. *Science 308*(5725):1149–54.

Christensen LA, Finch RA, Booker AJ, Vasquez KM. 2006. Targeting oncogenes to improve breast cancer chemotherapy. *Cancer Res 66*(8):4089–94.

Citron M, Vigo-Pelfrey C, Teplow DB, Miller C, Schenk D, Johnston J, Winblad B, Venizelos N, Lannfelt L, Selkoe DJ. 1994. Excessive production of amyloid betaprotein by peripheral cells of symptomatic and presymptomatic patients carrying the Swedish familial Alzheimer disease mutation. *Proc Natl Acad Sci U S A 91*(25):11993–7.

Clevers H. 2006. Colon cancer—understanding how NSAIDs work. *N Engl J Med 354*(7):761–3.

Cocheme HM, Kelso GF, James AM, Ross MF, Trnka J, Mahendiran T, Asin-Cayuela J, Blaikie FH, Manas AR, Porteous CM, and others. 2007. Mitochondrial targeting of quinones: Therapeutic implications. *Mitochondrion 7* Suppl 1:S94–S102.

Coffee B, Zhang F, Ceman S, Warren ST, Reines D. 2002. Histone modifications depict an aberrantly heterochromatinized FMR1 gene in fragile x syndrome. *Am J Hum Genet 71*(4):923–32.

Cohen JC, Kiss RS, Pertsemlidis A, Marcel YL, McPherson R, Hobbs HH. 2004. Multiple rare alleles contribute to low plasma levels of HDL cholesterol. *Science 305*(5685):869–72.

Cohn RD, van Erp C, Habashi JP, Soleimani AA, Klein EC, Lisi MT, Gamradt M, ap Rhys CM, Holm TM, Loeys BL, and others. 2007. Angiotensin II type 1 receptor blockade attenuates TGF-beta-induced failure of muscle regeneration in multiple myopathic states. *Nat Med 13*(2):204–10.

Conlin A, Smith G, Carey FA, Wolf CR, Steele RJ. 2005. The prognostic significance of K-ras, p53, and APC mutations in colorectal carcinoma. *Gut 54*(9):1283–6.

Conrad DF, Andrews TD, Carter NP, Hurles ME, Pritchard JK. 2006. A high-resolution survey of deletion polymorphism in the human genome. *Nat Genet 38*(1):75–81.

Coon KD, Myers AJ, Craig DW, Webster JA, Pearson JV, Lince DH, Zismann VL, Beach TG, Leung D, Bryden L, and others. 2007. A high-density whole-genome association study reveals that APOE is the major susceptibility gene for sporadic late-onset Alzheimer's disease. *J Clin Psychiatry 68*(4):613–8.

Coppede F, Mancuso M, Lo Gerfo A, Carlesi C, Piazza S, Rocchi A, Petrozzi L, Nesti C, Micheli D, Bacci A, and others. 2007. Association of the hOGG1 Ser326Cys polymorphism with sporadic amyotrophic lateral sclerosis. *Neurosci Lett 420*(2):163–8.

Couzin J, Kaiser J. 2005. Gene therapy. As Gelsinger case ends, gene therapy suffers another blow. *Science 307*(5712):1028.

Cox T, Lachmann R, Hollak C, Aerts J, van Weely S, Hrebicek M, Platt F, Butters T, Dwek R, Moyses C, and others. 2000. Novel oral treatment of Gaucher's disease with N-butyldeoxynojirimycin (OGT 918) to decrease substrate biosynthesis. *Lancet 355*(9214):1481–5.

Crabbe L, Jauch A, Naeger CM, Holtgreve-Grez H, Karlseder J. 2007. Telomere dysfunction as a cause of genomic instability in Werner syndrome. *Proc Natl Acad Sci U S A 104*(7):2205–10.

Cucca F, Lampis R, Congia M, Angius E, Nutland S, Bain SC, Barnett AH, Todd JA. 2001. A correlation between the relative predisposition of MHC class II alleles to type 1 diabetes and the structure of their proteins. *Hum Mol Genet 10*(19):2025–37.

Cui L, Jeong H, Borovecki F, Parkhurst CN, Tanese N, Krainc D. 2006. Transcriptional repression of PGC-1alpha by mutant huntingtin leads to mitochondrial dysfunction and neurodegeneration. *Cell 127*(1):59–69.

Cummings CJ, Zoghbi HY. 2000. Trinucleotide repeats: Mechanisms and pathophysiology. *Annu Rev Genomics Hum Genet 1*:281–328.

Dailey L, Laplantine E, Priore R, Basilico C. 2003. A network of transcriptional and signaling events is activated by FGF to induce chondrocyte growth arrest and differentiation. *J Cell Biol 161*(6):1053–66.

Damcott CM, Pollin TI, Reinhart LJ, Ott SH, Shen H, Silver KD, Mitchell BD, Shuldiner AR. 2006. Polymorphisms in the transcription factor 7-like 2 (TCF7L2) gene are associated with type 2 diabetes in the Amish: replication and evidence for a role in both insulin secretion and insulin resistance. *Diabetes 55*(9):2654–9.

Dan HC, Sun M, Yang L, Feldman RI, Sui XM, Ou CC, Nellist M, Yeung RS, Halley DJ, Nicosia SV, and others. 2002. Phosphatidylinositol 3-kinase/Akt pathway regulates tuberous sclerosis tumor suppressor complex by phosphorylation of tuberin. *J Biol Chem 277*(38):35364–70.

Dasgupta B, Gutmann DH. 2005. Neurofibromin regulates neural stem cell proliferation, survival, and astroglial differentiation in vitro and in vivo. *J Neurosci 25*(23): 5584–94.

Dave SS, Fu K, Wright GW, Lam LT, Kluin P, Boerma EJ, Greiner TC, Weisenburger DD, Rosenwald A, Ott G, and others. 2006. Molecular diagnosis of Burkitt's lymphoma. *N Engl J Med 354*(23):2431–42.

Deardorff MA, Kaur M, Yaeger D, Rampuria A, Korolev S, Pie J, Gil-Rodriguez C, Arnedo M, Loeys B, Kline AD, and others. 2007. Mutations in cohesin complex members SMC3 and SMC1A cause a mild variant of cornelia de Lange syndrome with predominant mental retardation. *Am J Hum Genet 80*(3):485–94.

Desai AA, Innocenti F, Ratain MJ. 2003. UGT pharmacogenomics: Implications for cancer risk and cancer therapeutics. *Pharmacogenetics 13*(8):517–23.

de Vries BB, Pfundt R, Leisink M, Koolen DA, Vissers LE, Janssen IM, Reijmersdal S, Nillesen WM, Huys EH, Leeuw N, and others. 2005. Diagnostic genome profiling in mental retardation. *Am J Hum Genet 77*(4):606–16.

Dietz HC, Pyeritz RE. 1995. Mutations in the human gene for fibrillin-1 (FBN1) in the Marfan syndrome and related disorders. *Hum Mol Genet 4* Spec No:1799–809.

Dompierre JP, Godin JD, Charrin BC, Cordelieres FP, King SJ, Humbert S, Saudou F. 2007. Histone deacetylase 6 inhibition compensates for the transport deficit in Huntington's disease by increasing tubulin acetylation. *J Neurosci 27*(13):3571–83.

Donnai D. 2002. Genetic services. *Clin Genet 61*(1):1–6.

Dorsett D, Eissenberg JC, Misulovin Z, Martens A, Redding B, McKim K. 2005. Effects of sister chromatid cohesion proteins on cut gene expression during wing development in Drosophila. *Development 132*(21):4743–53.

Drummond DC, Noble CO, Kirpotin DB, Guo Z, Scott GK, Benz CC. 2005. Clinical development of histone deacetylase inhibitors as anticancer agents. *Annu Rev Pharmacol Toxicol 45*:495–528.

Duan J, Wainwright MS, Comeron JM, Saitou N, Sanders AR, Gelernter J, Gejman PV. 2003. Synonymous mutations in the human dopamine receptor D2 (DRD2) affect mRNA stability and synthesis of the receptor. *Hum Mol Genet 12*(3): 205–16.

Duerr RH, Taylor KD, Brant SR, Rioux JD, Silverberg MS, Daly MJ, Steinhart AH, Abraham C, Regueiro M, Griffiths A, and others. 2006. A genome-wide association study identifies IL23R as an inflammatory bowel disease gene. *Science 314*(5804):1461–3.

Durand CM, Betancur C, Boeckers TM, Bockmann J, Chaste P, Fauchereau F, Nygren G, Rastam M, Gillberg IC, Anckarsater H, and others. 2007. Mutations in the gene encoding the synaptic scaffolding protein SHANK3 are associated with autism spectrum disorders. *Nat Genet 39*(1):25–7.

Dutta PR, Maity A. 2007. Cellular responses to EGFR inhibitors and their relevance to cancer therapy. *Cancer Lett 254*(2):165–77.

Dykxhoorn DM, Lieberman J. 2006. Knocking down disease with siRNAs. *Cell 126*(2):231–5.

Eckenhoff RG, Johansson JS, Wei H, Carnini A, Kang B, Wei W, Pidikiti R, Keller JM, Eckenhoff MF. 2004. Inhaled anesthetic enhancement of amyloid-beta oligomerization and cytotoxicity. *Anesthesiology 101*(3):703–9.

Eichler EE. 2006. Widening the spectrum of human genetic variation. *Nat Genet 38*(1):9–11.

Eldibany MM, Caprini JA. 2007. Hyperhomocysteinemia and thrombosis: An overview. *Arch Pathol Lab Med 131*(6):872–84.

El-Hashemite N, Zhang H, Henske EP, Kwiatkowski DJ. 2003. Mutation in TSC2 and activation of mammalian target of rapamycin signalling pathway in renal angiomyolipoma. *Lancet 361*(9366):1348–9.

Epeldegui M, Hung YP, McQuay A, Ambinder RF, Martinez-Maza O. 2007. Infection of human B cells with Epstein-Barr virus results in the expression of somatic hypermutation-inducing molecules and in the accrual of oncogene mutations. *Mol Immunol* 44(5):934–42.

Eriksen JL, Sagi SA, Smith TE, Weggen S, Das P, McLendon DC, Ozols VV, Jessing KW, Zavitz KH, Koo EH, and others. 2003. NSAIDs and enantiomers of flurbi-profen target gamma-secretase and lower Abeta 42 in vivo. *J Clin Invest* 112(3): 440–9.

Evans WE, Hon YY, Bomgaars L, Coutre S, Holdsworth M, Janco R, Kalwinsky D, Keller F, Khatib Z, Margolin J, and others. 2001. Preponderance of thiopurine S-methyltransferase deficiency and heterozygosity among patients intolerant to mercaptopurine or azathioprine. *J Clin Oncol* 19(8):2293–301.

Fan JQ, Ishii S, Asano N, Suzuki Y. 1999. Accelerated transport and maturation of lysosomal alpha-galactosidase A in Fabry lymphoblasts by an enzyme inhibitor. *Nat Med* 5(1):112–5.

Fantin VR, Leder P. 2006. Mitochondriotoxic compounds for cancer therapy. *Oncogene* 25(34):4787–97.

Fantin VR, St.-Pierre J, Leder P. 2006. Attenuation of LDH-A expression uncovers a link between glycolysis, mitochondrial physiology, and tumor maintenance. *Cancer Cell* 9(6):425–34.

Farmer H, McCabe N, Lord CJ, Tutt AN, Johnson DA, Richardson TB, Santarosa M, Dillon KJ, Hickson I, Knights C, and others. 2005. Targeting the DNA repair defect in BRCA mutant cells as a therapeutic strategy. *Nature* 434(7035):917–21.

Farrer LA, Cupples LA, Haines JL, Hyman B, Kukull WA, Mayeux R, Myers RH, Pericak-Vance MA, Risch N, van Duijn CM. 1997. Effects of age, sex, and ethnicity on the association between apolipoprotein E genotype and Alzheimer disease. A meta-analysis. APOE and Alzheimer Disease Meta Analysis Consortium. *JAMA* 278(16):1349–56.

Farris W, Schutz SG, Cirrito JR, Shankar GM, Sun X, George A, Leissring MA, Walsh DM, Qiu WQ, Holtzman DM, and others. 2007. Loss of neprilysin function promotes amyloid plaque formation and causes cerebral amyloid angiopathy. *Am J Pathol* 171(1):241–51.

Faustino NA, Cooper TA. 2003. Pre-mRNA splicing and human disease. *Genes Dev* 17(4):419–37.

Fears R, Weatherall D, Poste G. 1999. The impact of genetics on medical education and training. *Br Med Bull* 55(2):460–70.

Feenstra I, Brunner HG, van Ravenswaaij CM. 2006. Cytogenetic genotype-phenotype studies: Improving genotyping, phenotyping and data storage. *Cytogenet Genome Res* 115(3–4):231–9.

Feng Z, Hu W, de Stanchina E, Teresky AK, Jin S, Lowe S, Levine AJ. 2007. The regulation of AMPK beta1, TSC2, and PTEN expression by p53: Stress, cell and tissue specificity, and the role of these gene products in modulating the IGF-1-AKT-mTOR pathways. *Cancer Res* 67(7):3043–53.

Feuk L, Carson AR, Scherer SW. 2006. Structural variation in the human genome. *Nat Rev Genet* 7(2):85–97.

Fiegler H, Gribble SM, Burford DC, Carr P, Prigmore E, Porter KM, Clegg S, Crolla JA, Dennis NR, Jacobs P, and others. 2003. Array painting: a method for the rapid analysis of aberrant chromosomes using DNA microarrays. *J Med Genet 40*(9): 664–70.

Finck BN, Kelly DP. 2006. PGC-1 coactivators: inducible regulators of energy metabolism in health and disease. *J Clin Invest 116*(3):615–22.

Fire A, Xu S, Montgomery MK, Kostas SA, Driver SE, Mello CC. 1998. Potent and specific genetic interference by double-stranded RNA in *Caenorhabditis elegans. Nature 391*(6669):806–11.

Fischbach W, Goebeler-Kolve M, Starostik P, Greiner A, Muller-Hermelink HK. 2002. Minimal residual low-grade gastric MALT-type lymphoma after eradication of Helicobacter pylori. *Lancet 360*(9332):547–8.

Fishel R. 1999. Signaling mismatch repair in cancer. *Nat Med 5*(11):1239–41.

Fisher RA. 1918. The correlation between relatives on the supposition of Mendelian inheritance. *Trans R Soc Edinburgh 52*:399–433.

Fletcher JA. 2004. Role of KIT and platelet-derived growth factor receptors as oncoproteins. *Semin Oncol 31*(2 Suppl 6):4–11.

Franco RF, Reitsma PH. 2001. Genetic risk factors of venous thrombosis. *Hum Genet 109*(4):369–84.

Franz DN, Leonard J, Tudor C, Chuck G, Care M, Sethuraman G, Dinopoulos A, Thomas G, Crone KR. 2006. Rapamycin causes regression of astrocytomas in tuberous sclerosis complex. *Ann Neurol 59*(3):490–8.

Freedman DA, Levine AJ. 1999. Regulation of the p53 protein by the MDM2 oncoprotein—thirty-eighth G.H.A. Clowes Memorial Award Lecture. *Cancer Res 59*(1):1–7.

Freimer N, Sabatti C. 2003. The human phenome project. *Nat Genet 34*(1):15–21.

Fridlyand J, Snijders AM, Ylstra B, Li H, Olshen A, Segraves R, Dairkee S, Tokuyasu T, Ljung BM, Jain AN, and others. 2006. Breast tumor copy number aberration phenotypes and genomic instability. *BMC Cancer 6*:96.

Friedberg EC. 2003. DNA damage and repair. *Nature 421*(6921):436–40.

Fruehauf JP, Meyskens FL Jr. 2007. Reactive oxygen species: A breath of life or death? *Clin Cancer Res 13*(3):789–94.

Fuchs J, Nilsson C, Kachergus J, Munz M, Larsson EM, Schule B, Langston JW, Middleton FA, Ross OA, Hulihan M, and others. 2007. Phenotypic variation in a large Swedish pedigree due to SNCA duplication and triplication. *Neurology 68*(12):916–22.

Gallin J. 2002. *Principles and Practice of Clinical Research.* New York: Academic Press.

Gatz M, Reynolds CA, Fratiglioni L, Johansson B, Mortimer JA, Berg S, Fiske A, Pedersen NL. 2006. Role of genes and environments for explaining Alzheimer disease. *Arch Gen Psychiatry 63*(2):168–74.

Gaya DR, Russell RK, Nimmo ER, Satsangi J. 2006. New genes in inflammatory bowel disease: Lessons for complex diseases? *Lancet 367*(9518):1271–84.

Gerner EW, Meyskens FL Jr, Goldschmid S, Lance P, Pelot D. 2007. Rationale for, and design of, a clinical trial targeting polyamine metabolism for colon cancer chemoprevention. *Amino Acids* Mar 30; Epub ahead of print.

Giasson BI, Duda JE, Murray IV, Chen Q, Souza JM, Hurtig HI, Ischiropoulos H, Trojanowski JQ, Lee VM. 2000. Oxidative damage linked to neurodegeneration by selective alpha-synuclein nitration in synucleinopathy lesions. *Science 290*(5493): 985–9.

Gibson GL, Allsop D, Austen BM. 2004. Induction of cellular oxidative stress by the beta-amyloid peptide involved in Alzheimer's disease. *Protein Pept Lett 11*(3):257–70.

Gibson J, Morton NE, Collins A. 2006. Extended tracts of homozygosity in outbred human populations. *Hum Mol Genet 15*(5):789–95.

Gillis LA, McCallum J, Kaur M, DeScipio C, Yaeger D, Mariani A, Kline AD, Li HH, Devoto M, Jackson LG, and others. 2004. NIPBL mutational analysis in 120 individuals with Cornelia de Lange syndrome and evaluation of genotype-phenotype correlations. *Am J Hum Genet 75*(4):610–23.

Gius D, Spitz DR. 2006. Redox signaling in cancer biology. *Antioxid Redox Signal 8*(7–8):1249–52.

Godard B, Kaariainen H, Kristoffersson U, Tranebjaerg L, Coviello D, Ayme S. 2003. Provision of genetic services in Europe: Current practices and issues. *Eur J Hum Genet* Dec 11:S13–S48.

Goidts V, Armengol L, Schempp W, Conroy J, Nowak N, Muller S, Cooper DN, Estivill X, Enard W, Szamalek JM, and others. 2006. Identification of large-scale human-specific copy number differences by inter-species array comparative genomic hybridization. *Hum Genet 119*(1–2):185–98.

Goker-Alpan O, Hruska KS, Orvisky E, Kishnani PS, Stubblefield BK, Schiffmann R, Sidransky E. 2005. Divergent phenotypes in Gaucher disease implicate the role of modifiers. *J Med Genet 42*(6):e37.

Gold B, Merriam JE, Zernant J, Hancox LS, Taiber AJ, Gehrs K, Cramer K, Neel J, Bergeron J, Barile GR, and others. 2006. Variation in factor B (BF) and complement component 2 (C2) genes is associated with age-related macular degeneration. *Nat Genet 38*(4):458–62.

Goldberg AD, Allis CD, Bernstein E. 2007. Epigenetics: A landscape takes shape. *Cell 128*(4):635–8.

Goldstein DB, Cavalleri GL. 2005. Genomics: Understanding human diversity. *Nature 437*(7063):1241–2.

Gonzalez E, Kulkarni H, Bolivar H, Mangano A, Sanchez R, Catano G, Nibbs RJ, Freedman BI, Quinones MP, Bamshad MJ, and others. 2005. The influence of CCL3L1 gene-containing segmental duplications on HIV-1/AIDS susceptibility. *Science 307*(5714):1434–40.

Gracy RW, Talent JM, Kong Y, Conrad CC. 1999. Reactive oxygen species: The unavoidable environmental insult? *Mutat Res 428*(1–2):17–22.

Grant SF, Thorleifsson G, Reynisdottir I, Benediktsson R, Manolescu A, Sainz J, Helgason A, Stefansson H, Emilsson V, Helgadottir A, and others. 2006. Variant of transcription factor 7-like 2 (TCF7L2) gene confers risk of type 2 diabetes. *Nat Genet 38*(3):320–3.

Grantham R. 1974. Amino acid difference formula to help explain protein evolution. *Science 185*(4154):862–4.

Greene JG. 2006. Gene expression profiles of brain dopamine neurons and relevance to neuropsychiatric disease. *J Physiol 575*(Pt 2):411–6.

Greenman C, Stephens P, Smith R, Dalgliesh GL, Hunter C, Bignell G, Davies H, Teague J, Butler A, Stevens C, and others. 2007. Patterns of somatic mutation in human cancer genomes. *Nature 446*(7132):153–8.

Gregersen N. 2006. Protein misfolding disorders: Pathogenesis and intervention. *J Inherit Metab Dis 29*(2–3):456–70.

Gribble SM, Kalaitzopoulos D, Burford DC, Prigmore E, Selzer RR, Ng BL, Matthews NS, Porter KM, Curley R, Lindsay SJ, and others. 2007. Ultra-high resolution array painting facilitates breakpoint sequencing. *J Med Genet 44*(1): 51–8.

Guigo R, Valcarcel J. 2006. Unweaving the meanings of messenger RNA sequences. *Mol Cell 23*(2):150–1.

Gunderson KL, Steemers FJ, Lee G, Mendoza LG, Chee MS. 2005. A genome-wide scalable SNP genotyping assay using microarray technology. *Nat Genet 37*(5): 549–54.

Gupta GP, Nguyen DX, Chiang AC, Bos PD, Kim JY, Nadal C, Gomis RR, Manova-Todorova K, Massague J. 2007. Mediators of vascular remodelling co-opted for sequential steps in lung metastasis. *Nature 446*(7137):765–70.

Gutmann DH, Collins FS. 1993. The neurofibromatosis type 1 gene and its protein product, neurofibromin. *Neuron 10*(3):335–43.

Guymer R, Robman L. 2007. Chlamydia pneumoniae and age-related macular degeneration: A role in pathogenesis or merely a chance association? *Clin Experiment Ophthalmol 35*(1):89–93.

Habashi JP, Judge DP, Holm TM, Cohn RD, Loeys BL, Cooper TK, Myers L, Klein EC, Liu G, Calvi C, and others. 2006. Losartan, an AT1 antagonist, prevents aortic aneurysm in a mouse model of Marfan syndrome. *Science 312*(5770):117–21.

Haber DA, Settleman J. 2007. Cancer: Drivers and passengers. *Nature 446*(7132): 145–6.

Hafler DA, Compston A, Sawcer S, Lander ES, Daly MJ, De Jager PL, de Bakker PI, Gabriel SB, Mirel DB, Ivinson AJ, and others. 2007. Risk alleles for multiple sclerosis identified by a genomewide study. *N Engl J Med 357*(9):851–62.

Haines JL, Hauser MA, Schmidt S, Scott WK, Olson LM, Gallins P, Spencer KL, Kwan SY, Noureddine M, Gilbert JR, and others. 2005. Complement factor H variant increases the risk of age-related macular degeneration. *Science 308*(5720):419–21.

Hakonarson H, Grant SF, Bradfield JP, Marchand L, Kim CE, Glessner JT, Grabs R, Casalunovo T, Taback SP, Frackelton EC, and others. 2007. A genome-wide association study identifies KIAA0350 as a type 1 diabetes gene. *Nature 448*(7153): 591–4.

Hall JG. 2003. A clinician's plea. *Nat Genet 33*(4):440–2.

Hampe J, Cuthbert A, Croucher PJ, Mirza MM, Mascheretti S, Fisher S, Frenzel H, King K, Hasselmeyer A, MacPherson AJ, and others. 2001. Association between insertion mutation in NOD2 gene and Crohn's disease in German and British populations. *Lancet 357*(9272):1925–8.

Haque T, Wilkie GM, Jones MM, Higgins CD, Urquhart G, Wingate P, Burns D, McAulay K, Turner M, Bellamy C, and others. 2007. Allogeneic cytotoxic T cell therapy for EBV-positive post transplant lymphoproliferative disease: Results of a phase II multicentre clinical trial. *Blood 110*(4):1123–31.

Hayashi ML, Rao BS, Seo JS, Choi HS, Dolan BM, Choi SY, Chattarji S, Tonegawa S. 2007. Inhibition of p21-activated kinase rescues symptoms of fragile X syndrome in mice. *Proc Natl Acad Sci U S A 104*(27):11489–94.

Hegde MR, Chong B, Blazo ME, Chin LH, Ward PA, Chintagumpala MM, Kim JY, Plon SE, Richards CS. 2005. A homozygous mutation in MSH6 causes Turcot syndrome. *Clin Cancer Res 11*(13):4689–93.

Heitman J, Movva NR, Hall MN. 1991. Targets for cell cycle arrest by the immu-nosuppressant rapamycin in yeast. *Science 253*(5022):905–9.

Henikoff S, Henikoff JG. 1992. Amino acid substitution matrices from protein blocks. *Proc Natl Acad Sci U S A 89*(22):10915–9.

Hennah W, Tomppo L, Hiekkalinna T, Palo OM, Kilpinen H, Ekelund J, Tuulio-Henriksson A, Silander K, Partonen T, Paunio T, and others. 2007. Families with the risk allele of DISC1 reveal a link between schizophrenia and another compo-nent of the same molecular pathway, NDE1. *Hum Mol Genet 16*(5):453–62.

Henriksson K, Olsson H, Kristoffersson U. 2004. The need for oncogenetic coun-selling. Ten years' experience of a regional oncogenetic clinic. *Acta Oncol 43*(7): 637–49.

Herman JG, Graff JR, Myohanen S, Nelkin BD, Baylin SB. 1996. Methylation-specific PCR: A novel PCR assay for methylation status of CpG islands. *Proc Natl Acad Sci U S A 93*(18):9821–6.

Herry I, Neukirch C, Debray MP, Mignon F, Crestani B. 2007. Dramatic effect of sirolimus on renal angiomyolipomas in a patient with tuberous sclerosis complex. *Eur J Intern Med 18*(1):76–7.

Hershko A, Ciechanover A. 1998. The ubiquitin system. *Annu Rev Biochem 67*:425–79.

Hims MM, el Ibrahim C, Leyne M, Mull J, Liu L, Lazaro C, Shetty RS, Gill S, Gusella JF, Reed R, and others. 2007. Therapeutic potential and mechanism of kinetin as a treatment for the human splicing disease familial dysautonomia. *J Mol Med 85*(2): 149–61.

Hims MM, Shetty RS, Pickel J, Mull J, Leyne M, Liu L, Gusella JF, Slaugenhaupt SA. 2007. A humanized IKBKAP transgenic mouse models a tissue-specific human splicing defect. *Genomics 90*(3):389–96.

Hinault MP, Ben-Zvi A, Goloubinoff P. 2006. Chaperones and proteases: Cellular fold-controlling factors of proteins in neurodegenerative diseases and aging. *J Mol Neurosci 30*(3):249–65.

Hirsch-Reinshagen V, Wellington CL. 2007. Cholesterol metabolism, apolipoprotein E, adenosine triphosphate-binding cassette transporters, and Alzheimer's disease. *Curr Opin Lipidol 18*(3):325–32.

Hodges AK, Li S, Maynard J, Parry L, Braverman R, Cheadle JP, DeClue JE, Sampson JR. 2001. Pathological mutations in TSC1 and TSC2 disrupt the inter-action between hamartin and tuberin. *Hum Mol Genet 10*(25):2899–905.

Holscher C, Gengler S, Gault VA, Harriott P, Mallot HA. 2007. Soluble beta-amyloid[25–35] reversibly impairs hippocampal synaptic plasticity and spatial learning. *Eur J Pharmacol 561*(1–3):85–90.

Hondares E, Mora O, Yubero P, Rodriguez de la Concepcion M, Iglesias R, Giralt M, Villarroya F. 2006. Thiazolidinediones and rexinoids induce peroxisome proliferator-activated receptor-coactivator (PGC)-1alpha gene transcription: An autoregulatory loop controls PGC-1alpha expression in adipocytes via peroxi-some proliferator-activated receptor-gamma coactivation. *Endocrinology 147*(6): 2829–38.

Horton WA, Hall JG, Hecht JT. 2007. Achondroplasia. *Lancet 370*(9582):162–72.

Houtgraaf JH, Versmissen J, van der Giessen WJ. 2006. A concise review of DNA damage checkpoints and repair in mammalian cells. *Cardiovasc Revasc Med 7*(3): 165–72.

Hsu YC, Chern JJ, Cai Y, Liu M, Choi KW. 2007. Drosophila TCTP is essential for growth and proliferation through regulation of dRheb GTPase. *Nature 445*(7129): 785–8.

Huber KM, Gallagher SM, Warren ST, Bear MF. 2002. Altered synaptic plasticity in a mouse model of fragile X mental retardation. *Proc Natl Acad Sci U S A 99*(11):7746–50.

Huffmeier U, Zenker M, Hoyer J, Fahsold R, Rauch A. 2006. A variable combination of features of Noonan syndrome and neurofibromatosis type I are caused by mutations in the NF1 gene. *Am J Med Genet A 140*(24):2749–56.

Hugot JP, Cezard JP, Colombel JF, Belaiche J, Almer S, Tysk C, Montague S, Gassull M, Christensen S, Finkel Y, and others. 2003. Clustering of Crohn's disease within affected sibships. *Eur J Hum Genet 11*(2):179–84.

Hugot JP, Chamaillard M, Zouali H, Lesage S, Cezard JP, Belaiche J, Almer S, Tysk C, O'Morain CA, Gassull M, and others. 2001. Association of NOD2 leucine-rich repeat variants with susceptibility to Crohn's disease. *Nature 411*(6837):599–603.

Hugot JP, Laurent-Puig P, Gower-Rousseau C, Olson JM, Lee JC, Beaugerie L, Naom I, Dupas JL, Van Gossum A, Orholm M, and others. 1996. Mapping of a susceptibility locus for Crohn's disease on chromosome 16. *Nature 379*(6568):821–3.

Hulleman JD, Mirzaei H, Guigard E, Taylor KL, Ray SS, Kay CM, Regnier FE, Rochet JC. 2007. Destabilization of DJ-1 by familial substitution and oxidative modifications: Implications for Parkinson's disease. *Biochemistry 46*(19):5776–89.

Hummel M, Bentink S, Berger H, Klapper W, Wessendorf S, Barth TF, Bernd HW, Cogliatti SB, Dierlamm J, Feller AC, and others. 2006. A biologic definition of Burkitt's lymphoma from transcriptional and genomic profiling. *N Engl J Med 354*(23):2419–30.

Hurley LH, Von Hoff DD, Siddiqui-Jain A, Yang D. 2006. Drug targeting of the c-MYC promoter to repress gene expression via a G-quadruplex silencer element. *Semin Oncol 33*(4):498–512.

Hutter B, Helms V, Paulsen M. 2006. Tandem repeats in the CpG islands of imprinted genes. *Genomics 88*(3):323–32.

Inoki K, Ouyang H, Li Y, Guan KL. 2005. Signaling by target of rapamycin proteins in cell growth control. *Microbiol Mol Biol Rev 69*(1):79–100.

International HapMap Consortium. 2005. A haplotype map of the human genome. *Nature 437*(7063):1299–320.

International Multiple Sclerosis Consortium. 2007. Risk alleles for multiple sclerosis identified by a genomewide study. *N Engl J Med 357*(9):851–62.

Ionov Y, Peinado MA, Malkhosyan S, Shibata D, Perucho M. 1993. Ubiquitous somatic mutations in simple repeated sequences reveal a new mechanism for colonic carcinogenesis. *Nature 363*(6429):558–61.

Ischiropoulos H, Beckman JS. 2003. Oxidative stress and nitration in neurodegeneration: Cause, effect, or association? *J Clin Invest 111*(2):163–9.

Iyengar SK, Song D, Klein BE, Klein R, Schick JH, Humphrey J, Millard C, Liptak R, Russo K, Jun G, and others. 2004. Dissection of genomewide-scan data in extended families reveals a major locus and oligogenic susceptibility for age-related macular degeneration. *Am J Hum Genet 74*(1):20–39.

Jaenisch R, Bird A. 2003. Epigenetic regulation of gene expression: How the genome integrates intrinsic and environmental signals. *Nat Genet 33* Suppl:245–54.

Jakobsson J, Ekstrom L, Inotsume N, Garle M, Lorentzon M, Ohlsson C, Roh HK, Carlstrom K, Rane A. 2006. Large differences in testosterone excretion in Korean and Swedish men are strongly associated with a UDP-glucuronosyl transferase 2B17 polymorphism. *J Clin Endocrinol Metab 91*(2):687–93.

Jamain S, Quach H, Betancur C, Rastam M, Colineaux C, Gillberg IC, Soderstrom H, Giros B, Leboyer M, Gillberg C, and others. 2003. Mutations of the X-linked genes encoding neuroligins NLGN3 and NLGN4 are associated with autism. *Nat Genet 34*(1):27–9.

Jenne DE, Reimann H, Nezu J, Friedel W, Loff S, Jeschke R, Muller O, Back W, Zimmer M. 1998. Peutz-Jeghers syndrome is caused by mutations in a novel serine threonine kinase. *Nat Genet 18*(1):38–43.

Jenner P. 2003. Oxidative stress in Parkinson's disease. *Ann Neurol 53* Suppl 3:S26–36; discussion S36–8.

Jenuwein T, Allis CD. 2001. Translating the histone code. *Science 293*(5532):1074–80.

Jiang H, Poirier MA, Liang Y, Pei Z, Weiskittel CE, Smith WW, DeFranco DB, Ross CA. 2006. Depletion of CBP is directly linked with cellular toxicity caused by mutant huntingtin. *Neurobiol Dis 23*(3):543–51.

Jones PA. 2002. DNA methylation and cancer. *Oncogene 21*(35):5358–60.

Judge DP, Biery NJ, Keene DR, Geubtner J, Myers L, Huso DL, Sakai LY, Dietz HC. 2004. Evidence for a critical contribution of haploinsufficiency in the complex pathogenesis of Marfan syndrome. *J Clin Invest 114*(2):172–81.

Kakavanos R, Hopwood JJ, Lang D, Meikle PJ, Brooks DA. 2006. Stabilising normal and mis-sense variant alpha-glucosidase. *FEBS Lett 580*(18):4365–70.

Kamranvar SA, Gruhne B, Szeles A, Masucci MG. 2007. Epstein-Barr virus promotes genomic instability in Burkitt's lymphoma. *Oncogene 26*(35):5115–23.

Kang MK, Park NH. 2007. Extension of cell life span using exogenous telomerase. *Methods Mol Biol 371*:151–65.

Kaplitt MG, Feigin A, Tang C, Fitzsimons HL, Mattis P, Lawlor PA, Bland RJ, Young D, Strybing K, Eidelberg D, and others. 2007. Safety and tolerability of gene therapy with an adeno-associated virus (AAV) borne GAD gene for Parkinson's disease: An open label, phase I trial. *Lancet 369*(9579):2097–105.

Katajisto P, Vallenius T, Vaahtomeri K, Ekman N, Udd L, Tiainen M, Makela TP. 2007. The LKB1 tumor suppressor kinase in human disease. *Biochim Biophys Acta 1775*(1):63–75.

Kawamoto T, Araki K, Sonoda E, Yamashita YM, Harada K, Kikuchi K, Masutani C, Hanaoka F, Nozaki K, Hashimoto N, and others. 2005. Dual roles for DNA polymerase eta in homologous DNA recombination and translesion DNA synthesis. *Mol Cell 20*(5):793–9.

Kawanishi S, Oikawa S, Murata M. 2005. Evaluation for safety of antioxidant chemopreventive agents. *Antioxid Redox Signal 7*(11–12):1728–39.

Kehrer-Sawatzki H, Kluwe L, Sandig C, Kohn M, Wimmer K, Krammer U, Peyrl A, Jenne DE, Hansmann I, Mautner VF. 2004. High frequency of mosaicism among patients with neurofibromatosis type 1 (NF1) with microdeletions caused by somatic recombination of the JJAZ1 gene. *Am J Hum Genet 75*(3):410–23.

Khandjian EW, Huot ME, Tremblay S, Davidovic L, Mazroui R, Bardoni B. 2004. Biochemical evidence for the association of fragile X mental retardation protein with brain polyribosomal ribonucleoparticles. *Proc Natl Acad Sci U S A 101*(36): 13357–62.

Kim DH, Rossi JJ. 2007. Strategies for silencing human disease using RNA interference. *Nat Rev Genet 8*(3):173–84.

Kim JW, Zhang YH, Zern MA, Rossi JJ, Wu J. 2007. Short hairpin RNA causes the methylation of transforming growth factor-beta receptor II promoter and silencing of the target gene in rat hepatic stellate cells. *Biochem Biophys Res Commun 359*(2):292–7.

Kishore S, Stamm S. 2006. The snoRNA HBII-52 regulates alternative splicing of the serotonin receptor 2C. *Science 311*(5758):230–2.

Kitada T, Pisani A, Porter DR, Yamaguchi H, Tscherter A, Martella G, Bonsi P, Zhang C, Pothos EN, Shen J. 2007. From the cover: Impaired dopamine release and synaptic plasticity in the striatum of PINK1-deficient mice. *Proc Natl Acad Sci U S A 104*(27):11441–6.

Kleefstra T, Brunner HG, Amiel J, Oudakker AR, Nillesen WM, Magee A, Genevieve D, Cormier-Daire V, van Esch H, Fryns JP, and others. 2006. Loss-of-function mutations in euchromatin histone methyl transferase 1 (EHMT1) cause the 9q34 subtelomeric deletion syndrome. *Am J Hum Genet 79*(2):370–7.

Kleefstra T, Koolen DA, Nillesen WM, de Leeuw N, Hamel BC, Veltman JA, Sistermans EA, van Bokhoven H, van Ravenswaay C, de Vries BB. 2006. Interstitial 2.2 Mb deletion at 9q34 in a patient with mental retardation but without classical features of the 9q subtelomeric deletion syndrome. *Am J Med Genet A 140*(6): 618–23.

Klein C, Schlossmacher MG. 2006. The genetics of Parkinson disease: Implications for neurological care. *Nat Clin Pract Neurol 2*(3):136–46.

Klein E, Kis LL, Klein G. 2007. Epstein-Barr virus infection in humans: From harmless to life endangering virus-lymphocyte interactions. *Oncogene 26*(9):1297–305.

Klein ML, Schultz DW, Edwards A, Matise TC, Rust K, Berselli CB, Trzupek K, Weleber RG, Ott J, Wirtz MK, and others. 1998. Age-related macular degeneration. Clinical features in a large family and linkage to chromosome 1q. *Arch Ophthalmol 116*(8):1082–8.

Kovtun IV, Liu Y, Bjoras M, Klungland A, Wilson SH, McMurray CT. 2007. OGG1 initiates age-dependent CAG trinucleotide expansion in somatic cells. *Nature 447*(7143):447–52.

Kuipers EJ, Sipponen P. 2006. Helicobacter pylori eradication for the prevention of gastric cancer. *Helicobacter 11* Suppl 1:52–7.

Kure S, Sato K, Fujii K, Aoki Y, Suzuki Y, Kato S, Matsubara Y. 2004. Wild-type phenylalanine hydroxylase activity is enhanced by tetrahydrobiopterin supplementation in vivo: An implication for therapeutic basis of tetrahydrobiopterin-responsive phenylalanine hydroxylase deficiency. *Mol Genet Metab 83*(1–2): 150–6.

Lazarou J, Pomeranz BH, Corey PN. 1998. Incidence of adverse drug reactions in hospitalized patients: a meta-analysis of prospective studies. *JAMA 279*(15): 1200–5.

Lee A, Li W, Xu K, Bogert BA, Su K, Gao FB. 2003. Control of dendritic development by the *Drosophila* fragile X-related gene involves the small GTPase Rac1. *Development 130*(22):5543–52.

Lee JA, Lupski JR. 2006. Genomic rearrangements and gene copy-number alterations as a cause of nervous system disorders. *Neuron 52*(1):103–21.

Li Y, Inoki K, Guan KL. 2004. Biochemical and functional characterizations of small GTPase Rheb and TSC2 GAP activity. *Mol Cell Biol 24*(18):7965–75.

Litvan I, Chesselet MF, Gasser T, Di Monte DA, Parker D Jr, Hagg T, Hardy J, Jenner P, Myers RH, Price D, and others. 2007. The etiopathogenesis of Parkinson disease and suggestions for future research. Part II. *J Neuropathol Exp Neurol 66*(5):329–36.

Litvan I, Halliday G, Hallett M, Goetz CG, Rocca W, Duyckaerts C, Ben-Shlomo Y, Dickson DW, Lang AE, Chesselet MF, and others. 2007. The etiopathogenesis of Parkinson disease and suggestions for future research. Part I. *J Neuropathol Exp Neurol 66*(4):251–7.

Loos HS, Wieczorek D, Wurtz RP, von der Malsburg C, Horsthemke B. 2003. Computer-based recognition of dysmorphic faces. *Eur J Hum Genet 11*(8):555–60.

Lopez-Lluch G, Hunt N, Jones B, Zhu M, Jamieson H, Hilmer S, Cascajo MV, Allard J, Ingram DK, Navas P, and others. 2006. Calorie restriction induces mitochondrial biogenesis and bioenergetic efficiency. *Proc Natl Acad Sci U S A 103*(6): 1768–73.

Lu AL, Li X, Gu Y, Wright PM, Chang DY. 2001. Repair of oxidative DNA damage: Mechanisms and functions. *Cell Biochem Biophys 35*(2):141–70.

Lubs HA. 1969. A marker X chromosome. *Am J Hum Genet 21*(3):231–44.

Luoma P, Melberg A, Rinne JO, Kaukonen JA, Nupponen NN, Chalmers RM, Oldfors A, Rautakorpi I, Peltonen L, Majamaa K, and others. 2004. Parkinsonism, premature menopause, and mitochondrial DNA polymerase gamma mutations: clinical and molecular genetic study. *Lancet 364*(9437):875–82.

Mahley RW, Huang Y. 2006. Apolipoprotein (apo) E4 and Alzheimer's disease: Unique conformational and biophysical properties of apoE4 can modulate neuropathology. *Acta Neurol Scand 185* Suppl:8–14.

Mahley RW, Weisgraber KH, Huang Y. 2006. Apolipoprotein E4: A causative factor and therapeutic target in neuropathology, including Alzheimer's disease. *Proc Natl Acad Sci U S A 103*(15):5644–51.

Majumder PK, Febbo PG, Bikoff R, Berger R, Xue Q, McMahon LM, Manola J, Brugarolas J, McDonnell TJ, Golub TR, and others. 2004. mTOR inhibition reverses Akt-dependent prostate intraepithelial neoplasia through regulation of apoptotic and HIF-1-dependent pathways. *Nat Med 10*(6):594–601.

Malik SG, Pieter N, Sudoyo H, Kadir A, Marzuki S. 2003. Prevalence of the mitochondrial DNA A1555G mutation in sensorineural deafness patients in island Southeast Asia. *J Hum Genet 48*(9):480–3.

Maller J, George S, Purcell S, Fagerness J, Altshuler D, Daly MJ, Seddon JM. 2006. Common variation in three genes, including a noncoding variant in CFH, strongly influences risk of age-related macular degeneration. *Nat Genet 38*(9):1055–9.

Mandal PK, Pettegrew JW, McKeag DW, Mandal R. 2006. Alzheimer's disease: Halothane induces Abeta peptide to oligomeric form—solution NMR studies. *Neurochem Res 31*(7):883–90.

Mandal PK, Williams JP, Mandal R. 2007. Molecular understanding of Abeta peptide interaction with isoflurane, propofol, and thiopental: NMR spectroscopic study. *Biochemistry 46*(3):762–71.

Mandel S, Maor G, Youdim MB. 2004. Iron and alpha-synuclein in the substantia nigra of MPTP-treated mice: Effect of neuroprotective drugs R-apomorphine and green tea polyphenol (−)-epigallocatechin-3-gallate. *J Mol Neurosci 24*(3):401–16.

Manning BD, Cantley LC. 2003. Rheb fills a GAP between TSC and TOR. *Trends Biochem Sci 28*(11):573–6.

Mannon PJ, Fuss IJ, Mayer L, Elson CO, Sandborn WJ, Present D, Dolin B, Goodman N, Groden C, Hornung RL, and others. 2004. Anti-interleukin-12 antibody for active Crohn's disease. *N Engl J Med 351*(20):2069–79.

Markowitz SD. 2007. Aspirin and colon cancer—targeting prevention? *N Engl J Med 356*(21):2195–8.

Martin CL, Ledbetter DH. 2007. Autism and cytogenetic abnormalities: Solving autism one chromosome at a time. *Curr Psychiatry Rep 9*(2):141–7.

Martin JP, Bell J. 1943. A pedigree of mental defect showing sex linkage. *J Neurol Psychiat 6*:154–7.

Marx J. 2004. Preventing Alzheimer's: A lifelong commitment. *Science 309*: 863–6.

Marx J. 2007. Biomedicine. Puzzling out the pains in the gut. *Science 315*(5808): 33–5.

Masseroli M, Galati O, Manzotti M, Gibert K, Pinciroli F. 2005. Inherited disorder phenotypes: Controlled annotation and statistical analysis for knowledge mining from gene lists. *BMC Bioinformatics 6* Suppl 4:S18.

Masseroli M, Pinciroli F. 2006. Using gene ontology and genomic controlled vocabularies to analyze high-throughput gene lists: Three tool comparison. *Comput Biol Med 36*(7–8):731–47.

Mathew CG. 2006. Fanconi anaemia genes and susceptibility to cancer. *Oncogene 25*(43):5875–84.

Matsuda J, Suzuki O, Oshima A, Yamamoto Y, Noguchi A, Takimoto K, Itoh M, Matsuzaki Y, Yasuda Y, Ogawa S, and others. 2003. Chemical chaperone therapy for brain pathology in G(M1)-gangliosidosis. *Proc Natl Acad Sci U S A 100*(26):15912–7.

Matsumoto Y, Marusawa H, Kinoshita K, Endo Y, Kou T, Morisawa T, Azuma T, Okazaki IM, Honjo T, Chiba T. 2007. *Helicobacter pylori* infection triggers aberrant expression of activation-induced cytidine deaminase in gastric epithelium. *Nat Med 13*(4):470–6.

Matsuzaki H, Dong S, Loi H, Di X, Liu G, Hubbell E, Law J, Berntsen T, Chadha M, Hui H, and others. 2004. Genotyping over 100,000 SNPs on a pair of oligonucleotide arrays. *Nat Methods 1*(2):109–11.

Maxwell P, Salnikow K. 2004. HIF-1: An oxygen and metal responsive transcription factor. *Cancer Biol Ther 3*(1):29–35.

McLaurin J, Kierstead ME, Brown ME, Hawkes CA, Lambermon MH, Phinney AL, Darabie AA, Cousins JE, French JE, Lan MF, and others. 2006. Cyclohexanehexol inhibitors of Abeta aggregation prevent and reverse Alzheimer phenotype in a mouse model. *Nat Med 12*(7):801–8.

Mead LJ, Jenkins MA, Young J, Royce SG, Smith L, St. John DJ, Macrae F, Giles GG, Hopper JL, Southey MC. 2007. Microsatellite instability markers for identifying early-onset colorectal cancers caused by germ-line mutations in DNA mismatch repair genes. *Clin Cancer Res 13*(10):2865–9.

Mehta P, Saggar AK. 2005. Ethnicity, equity, and access to genetic services—the UK perspective. *Ann Hum Biol* 32(2):204–10.

Melillo G. 2007. Targeting hypoxia cell signaling for cancer therapy. *Cancer Metastasis Rev* 26(2):341–52.

Mensink KA, Ketterling RP, Flynn HC, Knudson RA, Lindor NM, Heese BA, Spinner RJ, Babovic-Vuksanovic D. 2006. Connective tissue dysplasia in five new patients with NF1 microdeletions: Further expansion of phenotype and review of the literature. *J Med Genet* 43(2):e8.

Merry BJ. 2004. Oxidative stress and mitochondrial function with aging—the effects of calorie restriction. *Aging Cell* 3(1):7–12.

Miller JW, Urbinati CR, Teng-Umnuay P, Stenberg MG, Byrne BJ, Thornton CA, Swanson MS. 2000. Recruitment of human muscleblind proteins to (CUG)(n) expansions associated with myotonic dystrophy. *Embo J* 19(17):4439–48.

Milyavsky M, Tabach Y, Shats I, Erez N, Cohen Y, Tang X, Kalis M, Kogan I, Buganim Y, Goldfinger N, and others. 2005. Transcriptional programs following genetic alterations in p53, INK4A, and H-Ras genes along defined stages of malignant transformation. *Cancer Res* 65(11):4530–43.

Ming JE, Geiger E, James AC, Ciprero KL, Nimmakayalu M, Zhang Y, Huang A, Vaddi M, Rappaport E, Zackai EH, and others. 2006. Rapid detection of submicroscopic chromosomal rearrangements in children with multiple congenital anomalies using high density oligonucleotide arrays. *Hum Mutat* 27(5):467–73.

Mizuta I, Satake W, Nakabayashi Y, Ito C, Suzuki S, Momose Y, Nagai Y, Oka A, Inoko H, Fukae J, and others. 2006. Multiple candidate gene analysis identifies alpha-synuclein as a susceptibility gene for sporadic Parkinson's disease. *Hum Mol Genet* 15(7):1151–8.

Modell B, Darr A. 2002. Science and society: Genetic counselling and customary consanguineous marriage. *Nat Rev Genet* 3(3):225–9.

Moffatt MF, Kabesch M, Liang L, Dixon AL, Strachan D, Heath S, Depner M, von Berg A, Bufe A, Rietschel E, and others. 2007. Genetic variants regulating ORMDL3 expression contribute to the risk of childhood asthma. *Nature* 448(7152): 470–3.

Moller LB, Tumer Z, Lund C, Petersen C, Cole T, Hanusch R, Seidel J, Jensen LR, Horn N. 2000. Similar splice-site mutations of the ATP7A gene lead to different phenotypes: Classical Menkes disease or occipital horn syndrome. *Am J Hum Genet* 66(4):1211–20.

Montaner S. 2007. Akt/TSC/mTOR activation by the KSHV G protein-coupled receptor: Emerging insights into the molecular oncogenesis and treatment of Kaposi's sarcoma. *Cell Cycle* 6(4):438–43.

Morel A, Boisdron-Celle M, Fey L, Laine-Cessac P, Gamelin E. 2007. Identification of a novel mutation in the dihydropyrimidine dehydrogenase gene in a patient with a lethal outcome following 5-fluorouracil administration and the determination of its frequency in a population of 500 patients with colorectal carcinoma. *Clin Biochem* 40(1–2):11–7.

Motoshima H, Goldstein BJ, Igata M, Araki E. 2006. AMPK and cell proliferation—AMPK as a therapeutic target for atherosclerosis and cancer. *J Physiol* 574(Pt 1): 63–71.

Muilu J, Peltonen L, Litton JE. 2007. The federated database—a basis for biobank-based post-genome studies, integrating phenome and genome data from 600,000 twin pairs in Europe. *Eur J Hum Genet 15*(7):718–23.

Muller CI, Ruter B, Koeffler HP, Lubbert M. 2006. DNA hypermethylation of myeloid cells, a novel therapeutic target in MDS and AML. *Curr Pharm Biotechnol 7*(5):315–21.

Musio A, Selicorni A, Focarelli ML, Gervasini C, Milani D, Russo S, Vezzoni P, Larizza L. 2006. X-linked Cornelia de Lange syndrome owing to SMC1L1 mutations. *Nat Genet 38*(5):528–30.

Myers AJ, Kaleem M, Marlowe L, Pittman AM, Lees AJ, Fung HC, Duckworth J, Leung D, Gibson A, Morris CM, and others. 2005. The H1c haplotype at the MAPT locus is associated with Alzheimer's disease. *Hum Mol Genet 14*(16): 2399–404.

Nachury MV, Loktev AV, Zhang Q, Westlake CJ, Peranen J, Merdes A, Slusarski DC, Scheller RH, Bazan JF, Sheffield VC, and others. 2007. A core complex of BBS proteins cooperates with the GTPase Rab8 to promote ciliary membrane biogenesis. *Cell 129*(6):1201–13.

Nackley AG, Shabalina SA, Tchivileva IE, Satterfield K, Korchynskyi O, Makarov SS, Maixner W, Diatchenko L. 2006. Human catechol-O-methyltransferase haplotypes modulate protein expression by altering mRNA secondary structure. *Science 314*(5807):1930–3.

Nagar S, Remmel RP. 2006. Uridine diphosphoglucuronosyltransferase pharmacogenetics and cancer. *Oncogene 25*(11):1659–72.

Nakabeppu Y, Tsuchimoto D, Yamaguchi H, Sakumi K. 2007. Oxidative damage in nucleic acids and Parkinson's disease. *J Neurosci Res 85*(5):919–34.

Neptune ER, Frischmeyer PA, Arking DE, Myers L, Bunton TE, Gayraud B, Ramirez F, Sakai LY, Dietz HC. 2003. Dysregulation of TGF-beta activation contributes to pathogenesis in Marfan syndrome. *Nat Genet 33*(3):407–11.

Netea MG, Ferwerda G, de Jong DJ, Jansen T, Jacobs L, Kramer M, Naber TH, Drenth JP, Girardin SE, Kullberg BJ, and others. 2005. Nucleotide-binding oligomerization domain-2 modulates specific TLR pathways for the induction of cytokine release. *J Immunol 174*(10):6518–23.

Neufeld EF. 1991. Lysosomal storage diseases. *Annu Rev Biochem 60*:257–80.

Newman M, Musgrave FI, Lardelli M. 2007. Alzheimer disease: Amyloidogenesis, the presenilins and animal models. *Biochim Biophys Acta 1772*(3):285–97.

Ng CM, Cheng A, Myers LA, Martinez-Murillo F, Jie C, Bedja D, Gabrielson KL, Hausladen JM, Mecham RP, Judge DP, and others. 2004. TGF-beta-dependent pathogenesis of mitral valve prolapse in a mouse model of Marfan syndrome. *J Clin Invest 114*(11):1586–92.

Nheu TV, He H, Hirokawa Y, Tamaki K, Florin L, Schmitz ML, Suzuki-Takahashi I, Jorissen RN, Burgess AW, Nishimura S, and others. 2002. The K252a derivatives, inhibitors for the PAK/MLK kinase family selectively block the growth of RAS transformants. *Cancer J 8*(4):328–36.

Nielsen KB, Sorensen S, Cartegni L, Corydon TJ, Doktor TK, Schroeder LD, Reinert LS, Elpeleg O, Krainer AR, Gregersen N, and others. 2007. Seemingly neutral polymorphic variants may confer immunity to splicing-inactivating mutations:

A synonymous SNP in exon 5 of MCAD protects from deleterious mutations in a flanking exonic splicing enhancer. *Am J Hum Genet 80*(3):416–32.

Oberle I, Rousseau F, Heitz D, Kretz C, Devys D, Hanauer A, Boue J, Bertheas MF, Mandel JL. 1991. Instability of a 550-base pair DNA segment and abnormal methylation in fragile X syndrome. *Science 252*(5010):1097–102.

O'Connor OA, Heaney ML, Schwartz L, Richardson S, Willim R, MacGregor-Cortelli B, Curly T, Moskowitz C, Portlock C, Horwitz S, and others. 2006. Clinical experience with intravenous and oral formulations of the novel histone deacetylase inhibitor suberoylanilide hydroxamic acid in patients with advanced hematologic malignancies. *J Clin Oncol 24*(1):166–73.

O'Donnell WT, Warren ST. 2002. A decade of molecular studies of fragile X syndrome. *Annu Rev Neurosci 25*:315–38.

Opalinska JB, Gewirtz AM. 2002. Nucleic-acid therapeutics: Basic principles and recent applications. *Nat Rev Drug Discov 1*(7):503–14.

O'Rahilly S, Wareham NJ. 2006. Genetic variants and common diseases—better late than never. *N Engl J Med 355*(3):306–8.

Osler W. 1906. *Address: Chauvinism in Medicine Aequanimitas.* New York: McGraw Hill.

Oti M, Snel B, Huynen MA, Brunner HG. 2006. Predicting disease genes using protein-protein interactions. *J Med Genet 43*(8):691–8.

Pai SI, Lin YY, Macaes B, Meneshian A, Hung CF, Wu TC. 2006. Prospects of RNA interference therapy for cancer. *Gene Ther 13*(6):464–77.

Pandey UB, Nie Z, Batlevi Y, McCray BA, Ritson GP, Nedelsky NB, Schwartz SL, DiProspero NA, Knight MA, Schuldiner O, and others. 2007. HDAC6 rescues neurodegeneration and provides an essential link between autophagy and the UPS. *Nature 447*(7146):859–63.

Pao W, Miller VA, Politi KA, Riely GJ, Somwar R, Zakowski MF, Kris MG, Varmus H. 2005. Acquired resistance of lung adenocarcinomas to gefitinib or erlotinib is associated with a second mutation in the EGFR kinase domain. *PLoS Med 2*(3):e73.

Paolini GV, Shapland RH, van Hoorn WP, Mason JS, Hopkins AL. 2006. Global mapping of pharmacological space. *Nat Biotechnol 24*(7):805–15.

Park J, Chen L, Ratnashinge L, Sellers TA, Tanner JP, Lee JH, Dossett N, Lang N, Kadlubar FF, Ambrosone CB, and others. 2006. Deletion polymorphism of UDP-glucuronosyltransferase 2B17 and risk of prostate cancer in African American and Caucasian men. *Cancer Epidemiol Biomarkers Prev 15*(8):1473–8.

Pearson H. 2006. Genetics: What is a gene? *Nature 441*(7092):398–401.

Peltonen L. 2007. Old suspects found guilty. *N Engl J Med* Jul 29; Epub ahead of print.

Pembrey M. 2004. The Avon Longitudinal Study of Parents and Children (ALSPAC): A resource for genetic epidemiology. *Eur J Endocrinol 151* Suppl 3:U125–9.

Pericak-Vance MA, St. George-Hyslop PH, Gaskell PC Jr, Growdon J, Crain BJ, Hulette C, Gusella JF, Yamaoka L, Tanzi RE, Roses AD, and others. 1993. Linkage analysis in familial Alzheimer disease: Description of the Duke and Boston data sets. *Genet Epidemiol 10*(6):361–4.

Perouansky M. 2007. Liaisons dangereuses? General anaesthetics and long-term toxicity in the CNS. *Eur J Anaesthesiol 24*(2):107–15.

Phillips KA, Veenstra DL, Oren E, Lee JK, Sadee W. 2001. Potential role of pharmacogenomics in reducing adverse drug reactions: A systematic review. *JAMA* *286*(18):2270–9.

Pierce GF, Lillicrap D, Pipe SW, Vandendriessche T. 2007. Gene therapy, bioengineered clotting factors and novel technologies for hemophilia treatment. *J Thromb Haemost* *5*(5):901–6.

Poirier J, Davignon J, Bouthillier D, Kogan S, Bertrand P, Gauthier S. 1993. Apolipoprotein E polymorphism and Alzheimer's disease. *Lancet* *342*(8873):697–9.

Potter CJ, Pedraza LG, Xu T. 2002. Akt regulates growth by directly phosphorylating Tsc2. *Nat Cell Biol* *4*(9):658–65.

Powers MV, Workman P. 2006. Targeting of multiple signalling pathways by heat shock protein 90 molecular chaperone inhibitors. *Endocr Relat Cancer* *13* Suppl 1:S125–35.

Prandini P, Deutsch S, Lyle R, Gagnebin M, Delucinge Vivier C, Delorenzi M, Gehrig C, Descombes P, Sherman S, Dagna Bricarelli F, and others. 2007. Natural gene-expression variation in Down syndrome modulates the outcome of gene-dosage imbalance. *Am J Hum Genet* *81*(2):252–63.

Pritchard DI. 2005. Sourcing a chemical succession for cyclosporin from parasites and human pathogens. *Drug Discov Today* *10*(10):688–91.

Quigley EM, Flourie B. 2007. Probiotics and irritable bowel syndrome: A rationale for their use and an assessment of the evidence to date. *Neurogastroenterol Motil* *19*(3):166–72.

Radin NS. 1996. Treatment of Gaucher disease with an enzyme inhibitor. *Glycoconj J* *13*(2):153–7.

Radtke HB, Sebold CD, Allison C, Haidle JL, Schneider G. 2007. Neurofibromatosis Type 1 in genetic counseling practice: Recommendations of the National Society of Genetic Counselors. *J Genet Couns* *16*(4):387–407.

Rattner A, Nathans J. 2006. Macular degeneration: Recent advances and therapeutic opportunities. *Nat Rev Neurosci* *7*(11):860–72.

Reardon DA, Wen PY. 2006. Therapeutic advances in the treatment of glioblastoma: Rationale and potential role of targeted agents. *Oncologist* *11*(2):152–64.

Redon R, Ishikawa S, Fitch KR, Feuk L, Perry GH, Andrews TD, Fiegler H, Shapero MH, Carson AR, Chen W, and others. 2006. Global variation in copy number in the human genome. *Nature* *444*(7118):444–54.

Reich DE, Cargill M, Bolk S, Ireland J, Sabeti PC, Richter DJ, Lavery T, Kouyoumjian R, Farhadian SF, Ward R, and others. 2001. Linkage disequilibrium in the human genome. *Nature* *411*(6834):199–204.

Reiman EM. 2007. Linking brain imaging and genomics in the study of Alzheimer's disease and aging. *Ann N Y Acad Sci* *1097*:94–113.

Reiman EM, Webster JA, Myers AJ, Hardy J, Dunckley T, Zismann VL, Joshipura KD, Pearson JV, Hu-Lince D, Huentelman MJ, and others. 2007. GAB2 alleles modify Alzheimer's risk in APOE epsilon4 carriers. *Neuron* *54*(5):713–20.

Ricker JL, Chen Z, Yang XP, Pribluda VS, Swartz GM, Van Waes C. 2004. 2-Methoxyestradiol inhibits hypoxia-inducible factor 1alpha, tumor growth, and angiogenesis and augments paclitaxel efficacy in head and neck squamous cell carcinoma. *Clin Cancer Res* *10*(24):8665–73.

Rideout WM 3rd, Eggan K, Jaenisch R. 2001. Nuclear cloning and epigenetic repro-gramming of the genome. *Science* 293(5532):1093–8.

Riekse RG, Li G, Petrie EC, Leverenz JB, Vavrek D, Vuletic S, Albers JJ, Montine TJ, Lee VM, Lee M, and others. 2006. Effect of statins on Alzheimer's disease bio-markers in cerebrospinal fluid. *J Alzheimers Dis* 10(4):399–406.

Risch N, Merikangas K. 1996. The future of genetic studies of complex human diseases. *Science* 273(5281):1516–7.

Risinger MA, Groden J. 2004. Crosslinks and crosstalk: human cancer syndromes and DNA repair defects. *Cancer Cell* 6(6):539–45.

Riva P, Castorina P, Manoukian S, Dalpra L, Doneda L, Marini G, den Dunnen J, Larizza L. 1996. Characterization of a cytogenetic 17q11.2 deletion in an NF1 patient with a contiguous gene syndrome. *Hum Genet* 98(6):646–50.

Roberson ED, Scearce-Levie K, Palop JJ, Yan F, Cheng IH, Wu T, Gerstein H, Yu GQ, Mucke L. 2007. Reducing endogenous tau ameliorates amyloid beta-induced deficits in an Alzheimer's disease mouse model. *Science* 316(5825):750–4.

Rocchi A, Pellegrini S, Siciliano G, Murri L. 2003. Causative and susceptibility genes for Alzheimer's disease: A review. *Brain Res Bull* 61(1):1–24.

Rocha D, Gut I, Jeffreys AJ, Kwok PY, Brookes AJ, Chanock SJ. 2006. Seventh international meeting on single nucleotide polymorphism and complex genome analysis: "Ever bigger scans and an increasingly variable genome." *Hum Genet* 119(4):451–6.

Rognan D. 2007. Chemogenomic approaches to rational drug design. *Br J Pharmacol* 152(1):38–52.

Rohas LM, St.-Pierre J, Uldry M, Jager S, Handschin C, Spiegelman BM. 2007. A fundamental system of cellular energy homeostasis regulated by PGC-1alpha. *Proc Natl Acad Sci U S A* 104(19):7933–8.

Ropers HH. 2007. New perspectives for the elucidation of genetic disorders. *Am J Hum Genet* 81(2):199–207.

Rosenberg RN. 2006. Treating Alzheimer disease: Time matters. *Arch Neurol* 63(7): 926–8.

Rosner M, Freilinger A, Hengstschlager M. 2007. Akt regulates nuclear/cytoplasmic localization of tuberin. *Oncogene* 26(4):521–31.

Ross CA, Thompson LM. 2006. Transcription meets metabolism in neurodegenera-tion. *Nat Med* 12(11):1239–41.

Rovelet-Lecrux A, Hannequin D, Raux G, Le Meur N, Laquerriere A, Vital A, Du-manchin C, Feuillette S, Brice A, Vercelletto M, and others. 2006. APP locus duplication causes autosomal dominant early-onset Alzheimer disease with cere-bral amyloid angiopathy. *Nat Genet* 38(1):24–6.

Rubinstein WS. 2007. Hereditary breast cancer: Pathobiology, clinical translation, and potential for targeted cancer therapeutics. *Fam Cancer* Jul 12; Epub ahead of print.

Sabatini DM. 2006. mTOR and cancer: Insights into a complex relationship. *Nat Rev Cancer* 6(9):729–34.

Sagaert X, De Wolf-Peeters C, Noels H, Baens M. 2007. The pathogenesis of MALT lymphomas: Where do we stand? *Leukemia* 21(3):389–96.

Sampson JR. 2003. TSC1 and TSC2: Genes that are mutated in the human genetic disorder tuberous sclerosis. *Biochem Soc Trans* 31(Pt 3):592–6.

Sanchez-Cespedes M, Parrella P, Esteller M, Nomoto S, Trink B, Engles JM, Westra WH, Herman JG, Sidransky D. 2002. Inactivation of LKB1/STK11 is a common event in adenocarcinomas of the lung. *Cancer Res 62*(13):3659–62.

Sandilands A, Terron-Kwiatkowski A, Hull PR, O'Regan GM, Clayton TH, Watson RM, Carrick T, Evans AT, Liao H, Zhao Y, and others. 2007. Comprehensive analysis of the gene encoding filaggrin uncovers prevalent and rare mutations in ichthyosis vulgaris and atopic eczema. *Nat Genet 39*(5):650–4.

Saxena R, Gianinny L, Burtt NP, Lyssenko V, Giuducci C, Sjogren M, Florez JC, Almgren P, Isomaa B, Orho-Melander M, and others. 2006. Common single nucleotide polymorphisms in TCF7L2 are reproducibly associated with type 2 diabetes and reduce the insulin response to glucose in nondiabetic individuals. *Diabetes 55*(10):2890–5.

Schaeffer C, Bardoni B, Mandel JL, Ehresmann B, Ehresmann C, Moine H. 2001. The fragile X mental retardation protein binds specifically to its mRNA via a purine quartet motif. *EMBO J 20*(17):4803–13.

Schaumberg DA, Christen WG, Buring JE, Glynn RJ, Rifai N, Ridker PM. 2007. High-sensitivity C-reactive protein, other markers of inflammation, and the incidence of macular degeneration in women. *Arch Ophthalmol 125*(3):300–5.

Schmitz A, Famulok M. 2007. Chemical biology: Ignore the nonsense. *Nature 447*(7140):42–3.

Schouten JP, McElgunn CJ, Waaijer R, Zwijnenburg D, Diepvens F, Pals G. 2002. Relative quantification of 40 nucleic acid sequences by multiplex ligation-dependent probe amplification. *Nucleic Acids Res 30*(12):e57.

Sebat J, Lakshmi B, Malhotra D, Troge J, Lese-Martin C, Walsh T, Yamrom B, Yoon S, Krasnitz A, Kendall J, and others. 2007. Strong association of de novo copy number mutations with autism. *Science 316*(5823):445–9.

Segditsas S, Tomlinson I. 2006. Colorectal cancer and genetic alterations in the Wnt pathway. *Oncogene 25*(57):7531–7.

Selivanova G, Wiman KG. 2007. Reactivation of mutant p53: Molecular mechanisms and therapeutic potential. *Oncogene 26*(15):2243–54.

Selkoe DJ. 2002. Alzheimer's disease is a synaptic failure. *Science 298*(5594):789–91.

Sharp AJ, Hansen S, Selzer RR, Cheng Z, Regan R, Hurst JA, Stewart H, Price SM, Blair E, Hennekam RC, and others. 2006. Discovery of previously unidentified genomic disorders from the duplication architecture of the human genome. *Nat Genet 38*(9):1038–42.

Sharp SY, Prodromou C, Boxall K, Powers MV, Holmes JL, Box G, Matthews TP, Cheung KM, Kalusa A, James K, and others. 2007. Inhibition of the heat shock protein 90 molecular chaperone in vitro and in vivo by novel, synthetic, potent resorcinylic pyrazole/isoxazole amide analogues. *Mol Cancer Ther 6*(4):1198–211.

Shaw RJ. 2006. Glucose metabolism and cancer. *Curr Opin Cell Biol 18*(6):598–608.

Shaw-Smith C, Pittman AM, Willatt L, Martin H, Rickman L, Gribble S, Curley R, Cumming S, Dunn C, Kalaitzopoulos D, and others. 2006. Microdeletion encompassing MAPT at chromosome 17q21.3 is associated with developmental delay and learning disability. *Nat Genet 38*(9):1032–7.

Shelbourne PF, Keller-McGandy C, Bi WL, Yoon SR, Dubeau L, Veitch NJ, Vonsattel JP, Wexler NS, Arnheim N, Augood SJ. 2007. Triplet repeat mutation length gains correlate with cell-type specific vulnerability in Huntington disease brain. *Hum Mol Genet 16*(10):1133–42.

Sherer TB, Betarbet R, Testa CM, Seo BB, Richardson JR, Kim JH, Miller GW, Yagi T, Matsuno-Yagi A, Greenamyre JT. 2003. Mechanism of toxicity in rotenone models of Parkinson's disease. *J Neurosci 23*(34):10756–64.

Shillingford JM, Murcia NS, Larson CH, Low SH, Hedgepeth R, Brown N, Flask CA, Novick AC, Goldfarb DA, Kramer-Zucker A, and others. 2006. The mTOR pathway is regulated by polycystin-1, and its inhibition reverses renal cystogenesis in polycystic kidney disease. *Proc Natl Acad Sci U S A 103*(14): 5466–71.

Shuraih M, Ai T, Vatta M, Sohma Y, Merkle EM, Taylor E, Li Z, Xi Y, Razavi M, Towbin JA, and others. 2007. A common SCN5A variant alters the responsiveness of human sodium channels to class I antiarrhythmic agents. *J Cardiovasc Electrophysiol 18*(4):434–40.

Siemers ER, Quinn JF, Kaye J, Farlow MR, Porsteinsson A, Tariot P, Zoulnouni P, Galvin JE, Holtzman DM, Knopman DS, and others. 2006. Effects of a gamma-secretase inhibitor in a randomized study of patients with Alzheimer disease. *Neurology 66*(4):602–4.

Sillen A, Forsell C, Lilius L, Axelman K, Bjork BF, Onkamo P, Kere J, Winblad B, Graff C. 2006. Genome scan on Swedish Alzheimer's disease families. *Mol Psychiatry 11*(2):182–6.

Silverman JM, Li G, Zaccario ML, Smith CJ, Schmeidler J, Mohs RC, Davis KL. 1994. Patterns of risk in first-degree relatives of patients with Alzheimer's disease. *Arch Gen Psychiatry 51*(7):577–86.

Singleton AB, Farrer M, Johnson J, Singleton A, Hague S, Kachergus J, Hulihan M, Peuralinna T, Dutra A, Nussbaum R, and others. 2003. alpha-Synuclein locus triplication causes Parkinson's disease. *Science 302*(5646):841.

Sjoblom T, Jones S, Wood LD, Parsons DW, Lin J, Barber TD, Mandelker D, Leary RJ, Ptak J, Silliman N, and others. 2006. The consensus coding sequences of human breast and colorectal cancers. *Science 314*(5797):268–74.

Sladek R, Rocheleau G, Rung J, Dina C, Shen L, Serre D, Boutin P, Vincent D, Belisle A, Hadjadj S, and others. 2007. A genome-wide association study identifies novel risk loci for type 2 diabetes. *Nature 445*(7130):881–5.

Slaugenhaupt SA, Blumenfeld A, Gill SP, Leyne M, Mull J, Cuajungco MP, Liebert CB, Chadwick B, Idelson M, Reznik L, and others. 2001. Tissue-specific expression of a splicing mutation in the IKBKAP gene causes familial dysautonomia. *Am J Hum Genet 68*(3):598–605.

Slaugenhaupt SA, Mull J, Leyne M, Cuajungco MP, Gill SP, Hims MM, Quintero F, Axelrod FB, Gusella JF. 2004. Rescue of a human mRNA splicing defect by the plant cytokinin kinetin. *Hum Mol Genet 13*(4):429–36.

Smith DP, Ciccotosto GD, Tew DJ, Fodero-Tavoletti MT, Johanssen T, Masters CL, Barnham KJ, Cappai R. 2007. Concentration dependent Cu2+ induced aggregation and dityrosine formation of the Alzheimer's disease amyloid-beta peptide. *Biochemistry 46*(10):2881–91.

Smyth D, Cooper JD, Collins JE, Heward JM, Franklyn JA, Howson JM, Vella A, Nutland S, Rance HE, Maier L, and others. 2004. Replication of an association between the lymphoid tyrosine phosphatase locus (LYP/PTPN22) with type 1 diabetes, and evidence for its role as a general autoimmunity locus. *Diabetes* *53*(11):3020–3.

Sodhi A, Chaisuparat R, Hu J, Ramsdell AK, Manning BD, Sausville EA, Sawai ET, Molinolo A, Gutkind JS, Montaner S. 2006. The TSC2/mTOR pathway drives endothelial cell transformation induced by the Kaposi's sarcoma-associated herpesvirus G protein-coupled receptor. *Cancer Cell 10*(2):133–43.

Soret J, Bakkour N, Maire S, Durand S, Zekri L, Gabut M, Fic W, Divita G, Rivalle C, Dauzonne D, and others. 2005. Selective modification of alternative splicing by indole derivatives that target serine-arginine-rich protein splicing factors. *Proc Natl Acad Sci U S A 102*(24):8764–9.

Stefansson H, Helgason A, Thorleifsson G, Steinthorsdottir V, Masson G, Barnard J, Baker A, Jonasdottir A, Ingason A, Gudnadottir VG, and others. 2005. A common inversion under selection in Europeans. *Nat Genet 37*(2):129–37.

St.-Pierre J, Buckingham JA, Roebuck SJ, Brand MD. 2002. Topology of superoxide production from different sites in the mitochondrial electron transport chain. *J Biol Chem 277*(47):44784–90.

Strachan T, Read AP. 2003. *Human Molecular Genetics*, 3rd ed. London and New York: Garland Science Taylor and Francis Group.

Stranger BE, Forrest MS, Dunning M, Ingle CE, Beazley C, Thorne N, Redon R, Bird CP, de Grassi A, Lee C, and others. 2007. Relative impact of nucleotide and copy number variation on gene expression phenotypes. *Science 315*(5813):848–53.

Sugars KL, Rubinsztein DC. 2003. Transcriptional abnormalities in Huntington disease. *Trends Genet 19*(5):233–8.

Sugnet CW, Srinivasan K, Clark TA, O'Brien G, Cline MS, Wang H, Williams A, Kulp D, Blume JE, Haussler D, and others. 2006. Unusual intron conservation near tissue-regulated exons found by splicing microarrays. *PLoS Comput Biol 2*(1):e4.

Suzuki H, Hibi T, Marshall BJ. 2007. *Helicobacter pylori*: Present status and future prospects in Japan. *J Gastroenterol 42*(1):1–15.

Swinnen LJ. 2001. Post-transplant lymphoproliferative disorders: Implications for acquired immunodeficiency syndrome-associated malignancies. *J Natl Cancer Inst Monogr* (28):38–43.

Szatmari P, Paterson AD, Zwaigenbaum L, Roberts W, Brian J, Liu XQ, Vincent JB, Skaug JL, Thompson AP, Senman L, and others. 2007. Mapping autism risk loci using genetic linkage and chromosomal rearrangements. *Nat Genet 39*(3):319–28.

Taddei K, Fisher C, Laws SM, Martins G, Paton A, Clarnette RM, Chung C, Brooks WS, Hallmayer J, Miklossy J, and others. 2002. Association between presenilin-1 Glu318Gly mutation and familial Alzheimer's disease in the Australian population. *Mol Psychiatry 7*(7):776–81.

Taille C, Debray MP, Crestani B. 2007. Sirolimus treatment for pulmonary lymphangioleiomyomatosis. *Ann Intern Med 146*(9):687–8.

Tang J, Robertson S, Lem KE, Godwin SC, Kaler SG. 2006. Functional copper transport explains neurologic sparing in occipital horn syndrome. *Genet Med 8*(11):711–8.

Tantisira KG, Lake S, Silverman ES, Palmer LJ, Lazarus R, Silverman EK, Liggett SB, Gelfand EW, Rosenwasser LJ, Richter B, and others. 2004. Corticosteroid pharmacogenetics: Association of sequence variants in CRHR1 with improved lung function in asthmatics treated with inhaled corticosteroids. *Hum Mol Genet* *13*(13):1353–9.

Thomas L. 1978. *The Lives of a Cell.* New York: Penguin.

Tian G, Lai L, Guo H, Lin Y, Butchbach ME, Chang Y, Lin CL. 2007. Translational control of glial glutamate transporter EAAT2 expression. *J Biol Chem* *282*(3): 1727–37.

Todaro GJ, Fryling C, De Larco JE. 1980. Transforming growth factors produced by certain human tumor cells: Polypeptides that interact with epidermal growth factor receptors. *Proc Natl Acad Sci U S A* *77*(9):5258–62.

Tonkin ET, Smith M, Eichhorn P, Jones S, Imamwerdi B, Lindsay S, Jackson M, Wang TJ, Ireland M, Burn J, and others. 2004. A giant novel gene undergoing extensive alternative splicing is severed by a Cornelia de Lange-associated translocation breakpoint at 3q26.3. *Hum Genet* *115*(2):139–48.

Tonkin ET, Wang TJ, Lisgo S, Bamshad MJ, Strachan T. 2004. NIPBL, encoding a homolog of fungal Scc2-type sister chromatid cohesion proteins and fly Nipped-B, is mutated in Cornelia de Lange syndrome. *Nat Genet* *36*(6):636–41.

Trotman LC, Alimonti A, Scaglioni PP, Koutcher JA, Cordon-Cardo C, Pandolfi PP. 2006. Identification of a tumour suppressor network opposing nuclear Akt function. *Nature* *441*(7092):523–7.

Tuynder M, Fiucci G, Prieur S, Lespagnol A, Geant A, Beaucourt S, Duflaut D, Besse S, Susini L, Cavarelli J, and others. 2004. Translationally controlled tumor protein is a target of tumor reversion. *Proc Natl Acad Sci U S A* *101*(43):15364–9.

Tuynder M, Susini L, Prieur S, Besse S, Fiucci G, Amson R, Telerman A. 2002. Biological models and genes of tumor reversion: Cellular reprogramming through tpt1/TCTP and SIAH-1. *Proc Natl Acad Sci U S A* *99*(23):14976–81.

Ule J, Darnell RB. 2006. RNA binding proteins and the regulation of neuronal synaptic plasticity. *Curr Opin Neurobiol* *16*(1):102–10.

Ullmann R, Turner G, Kirchhoff M, Chen W, Tonge B, Rosenberg C, Field M, Vianna-Morgante AM, Christie L, Krepischi-Santos AC, and others. 2007. Array CGH identifies reciprocal 16p13.1 duplications and deletions that predispose to autism and/or mental retardation. *Hum Mutat* *28*(7):674–82.

Vallur AC, Yabuki M, Larson ED, Maizels N. 2007. AID in antibody perfection. *Cell Mol Life Sci* *64*(5):555–65.

Vance JE, Karten B, Hayashi H. 2006. Lipid dynamics in neurons. *Biochem Soc Trans* *34*(Pt 3):399–403.

van der Walt JM, Nicodemus KK, Martin ER, Scott WK, Nance MA, Watts RL, Hubble JP, Haines JL, Koller WC, Lyons K, and others. 2003. Mitochondrial polymorphisms significantly reduce the risk of Parkinson disease. *Am J Hum Genet* *72*(4):804–11.

van Driel MA, Bruggeman J, Vriend G, Brunner HG, Leunissen JA. 2006. A text-mining analysis of the human phenome. *Eur J Hum Genet* *14*(5):535–42.

van Kuilenburg AB, Meinsma R, Zonnenberg BA, Zoetekouw L, Baas F, Matsuda K, Tamaki N, van Gennip AH. 2003. Dihydropyrimidinase deficiency and severe 5-fluorouracil toxicity. *Clin Cancer Res* *9*(12):4363–7.

Van Patten SM, Hughes H, Huff MR, Piepenhagen PA, Waire J, Qiu H, Ganesa C, Reczek D, Ward PV, Kutzko JP, and others. 2007. Effect of mannose chain length on targeting of glucocerebrosidase for enzyme replacement therapy of Gaucher disease. *Glycobiology 17*(5):467–78.

Varmus H. 2006. The new era in cancer research. *Science 312*(5777):1162–5.

Vassilev LT, Vu BT, Graves B, Carvajal D, Podlaski F, Filipovic Z, Kong N, Kammlott U, Lukacs C, Klein C, and others. 2004. In vivo activation of the p53 pathway by small-molecule antagonists of MDM2. *Science 303*(5659):844–8.

Vega H, Waisfisz Q, Gordillo M, Sakai N, Yanagihara I, Yamada M, van Gosliga D, Kayserili H, Xu C, Ozono K, and others. 2005. Roberts syndrome is caused by mutations in ESCO2, a human homolog of yeast ECO1 that is essential for the establishment of sister chromatid cohesion. *Nat Genet 37*(5):468–70.

Vitkup D, Sander C, Church GM. 2003. The amino-acid mutational spectrum of human genetic disease. *Genome Biol 4*(11):R72.

Vogelstein B, Kinzler KW. 1992. p53 function and dysfunction. *Cell 70*(4):523–6.

Vogelstein B, Kinzler KW. 2004. Cancer genes and the pathways they control. *Nat Med 10*(8):789–99.

Vogelstein B, Lane D, Levine AJ. 2000. Surfing the p53 network. *Nature 408*(6810): 307–10.

Wall NR, Shi Y. 2003. Small RNA: Can RNA interference be exploited for therapy? *Lancet 362*(9393):1401–3.

Wallace DC. 2005. A mitochondrial paradigm of metabolic and degenerative diseases, aging, and cancer: A dawn for evolutionary medicine. *Annu Rev Genet 39*:359–407.

Walsh DM, Selkoe DJ. 2004. Deciphering the molecular basis of memory failure in Alzheimer's disease. *Neuron 44*(1):181–93.

Wang Q, Montmain G, Ruano E, Upadhyaya M, Dudley S, Liskay RM, Thibodeau SN, Puisieux A. 2003. Neurofibromatosis type 1 gene as a mutational target in a mismatch repair-deficient cell type. *Hum Genet 112*(2):117–23.

Warburg O. 1930. *The Metabolism of Tumors*. London: Arnold Constable.

Watanabe T, Kitani A, Murray PJ, Strober W. 2004. NOD2 is a negative regulator of Toll-like receptor 2-mediated T helper type 1 responses. *Nat Immunol 5*(8):800–8.

Watanabe T, Kitani A, Strober W. 2005. NOD2 regulation of Toll-like receptor responses and the pathogenesis of Crohn's disease. *Gut 54*(11):1515–8.

Waters MD, Olden K, Tennant RW. 2003. Toxicogenomic approach for assessing toxicant-related disease. *Mutat Res 544*(2–3):415–24.

Watson FL, Puttmann-Holgado R, Thomas F, Lamar DL, Hughes M, Kondo M, Rebel VI, Schmucker D. 2005. Extensive diversity of Ig-superfamily proteins in the immune system of insects. *Science 309*(5742):1874–8.

Weatherall DJ. 2005. The global problem of genetic disease. *Ann Hum Biol 32*(2): 117–22.

Weeks DE, Conley YP, Tsai HJ, Mah TS, Schmidt S, Postel EA, Agarwal A, Haines JL, Pericak-Vance MA, Rosenfeld PJ, and others. 2004. Age-related maculopathy: a genomewide scan with continued evidence of susceptibility loci within the 1q31, 10q26, and 17q25 regions. *Am J Hum Genet 75*(2):174–89.

Wehkamp J, Harder J, Weichenthal M, Schwab M, Schaffeler E, Schlee M, Herrlinger KR, Stallmach A, Noack F, Fritz P, and others. 2004. NOD2 (CARD15) mutations in Crohn's disease are associated with diminished mucosal alpha-defensin expression. *Gut 53*(11):1658–64.

Weihl CC, Connolly AM, Pestronk A. 2006. Valproate may improve strength and function in patients with type III/IV spinal muscle atrophy. *Neurology 67*(3):500–1.

Weinberg RA. 2007. *The Biology of Cancer*. New York: Garland Science.

Weissenbach J. 1998. The Human Genome Project: From mapping to sequencing. *Clin Chem Lab Med 36*(8):511–4.

Weisz L, Damalas A, Liontos M, Karakaidos P, Fontemaggi G, Maor-Aloni R, Kalis M, Levrero M, Strano S, Gorgoulis VG, and others. 2007. Mutant p53 enhances nuclear factor kappaB activation by tumor necrosis factor alpha in cancer cells. *Cancer Res 67*(6):2396–401.

Weisz L, Oren M, Rotter V. 2007. Transcription regulation by mutant p53. *Oncogene 26*(15):2202–11.

Welch EM, Barton ER, Zhuo J, Tomizawa Y, Friesen WJ, Trifillis P, Paushkin S, Patel M, Trotta CR, Hwang S, and others. 2007. PTC124 targets genetic disorders caused by nonsense mutations. *Nature 447*(7140):87–91.

Weydt P, Pineda VV, Torrence AE, Libby RT, Satterfield TF, Lazarowski ER, Gilbert ML, Morton GJ, Bammler TK, Strand AD, and others. 2006. Thermoregulatory and metabolic defects in Huntington's disease transgenic mice implicate PGC-1alpha in Huntington's disease neurodegeneration. *Cell Metab 4*(5):349–62.

Wienecke R, Fackler I, Linsenmaier U, Mayer K, Licht T, Kretzler M. 2006. Anti-tumoral activity of rapamycin in renal angiomyolipoma associated with tuberous sclerosis complex. *Am J Kidney Dis 48*(3):e27–9.

Wilquet V, De Strooper B. 2004. Amyloid-beta precursor protein processing in neurodegeneration. *Curr Opin Neurobiol 14*(5):582–8.

Wilson W 3rd, Pardo-Manuel de Villena F, Lyn-Cook BD, Chatterjee PK, Bell TA, Detwiler DA, Gilmore RC, Valladeras IC, Wright CC, Threadgill DW, and others. 2004. Characterization of a common deletion polymorphism of the UGT2B17 gene linked to UGT2B15. *Genomics 84*(4):707–14.

Wirth B, Brichta L, Hahnen E. 2006. Spinal muscular atrophy and therapeutic prospects. *Prog Mol Subcell Biol 44*:109–32.

Wu WS. 2006. The signaling mechanism of ROS in tumor progression. *Cancer Metastasis Rev 25*(4):695–705.

Wu X, Gu J, Spitz MR. 2007. Mutagen sensitivity: A genetic predisposition factor for cancer. *Cancer Res 67*(8):3493–5.

Xie Z, Dong Y, Maeda U, Moir RD, Xia W, Culley DJ, Crosby G, Tanzi RE. 2007. The inhalation anesthetic isoflurane induces a vicious cycle of apoptosis and amyloid beta-protein accumulation. *J Neurosci 27*(6):1247–54.

Yasui S, Tsuzaki K, Ninomiya H, Floricel F, Asano Y, Maki H, Takamura A, Nanba E, Higaki K, Ohno K. 2007. The TSC1 gene product hamartin interacts with NADE. *Mol Cell Neurosci 35*(1):100–8.

Ye S, Huang Y, Mullendorff K, Dong L, Giedt G, Meng EC, Cohen FE, Kuntz ID, Weisgraber KH, Mahley RW. 2005. Apolipoprotein (apo) E4 enhances amyloid beta peptide production in cultured neuronal cells: ApoE structure as a potential therapeutic target. *Proc Natl Acad Sci U S A 102*(51):18700–5.

Yilmaz OH, Valdez R, Theisen BK, Guo W, Ferguson DO, Wu H, Morrison SJ. 2006. Pten dependence distinguishes haematopoietic stem cells from leukaemia-initiating cells. *Nature 441*(7092):475–82.

Zhang J, Grindley JC, Yin T, Jayasinghe S, He XC, Ross JT, Haug JS, Rupp D, Porter-Westpfahl KS, Wiedemann LM, and others. 2006. PTEN maintains haemato-poietic stem cells and acts in lineage choice and leukaemia prevention. *Nature 441*(7092):518–22.

Zhou B, Teramukai S, Fukushima M. 2007. Prevention and treatment of dementia or Alzheimer's disease by statins: A meta-analysis. *Dement Geriatr Cogn Disord 23*(3):194–201.

Zhou W, Song W. 2006. Leaky scanning and reinitiation regulate BACE1 gene expression. *Mol Cell Biol 26*(9):3353–64.

Zhou W, Zhu M, Wilson MA, Petsko GA, Fink AL. 2006. The oxidation state of DJ-1 regulates its chaperone activity toward alpha-synuclein. *J Mol Biol 356*(4): 1036–48.

INDEX